Evolution of the
Metazoan Life Cycle

Evolution of the Metazoan Life Cycle

A comprehensive theory

GÖSTA JÄGERSTEN

Institute of Zoology,
University of Uppsala, Sweden

1972

Academic Press

London and New York

ACADEMIC PRESS INC. (LONDON) LTD.
24/28 Oval Road,
London NW1

United States Edition published by
ACADEMIC PRESS INC.
111 Fifth Avenue
New York, New York 10003

Library of Congress Catalog Card Number: 76–149700
ISBN: 0–12–379950–3

PRINTED IN GREAT BRITAIN BY
WILLIAM CLOWES & SONS, LIMITED
LONDON, BECCLES AND COLCHESTER

Preface

In the course of preparations for extending my earlier phylogenetic investigations (Jägersten, 1955, 1959) to all the metazoan groups, I frequently found it impossible to arrive at satisfactory conclusions because a comprehensive view of the evolution of the metazoan life cycle did not exist. Today, it is established that not only the adult but all stages of the life cycle are subjected to hereditary changes. What is not known, however, is whether these changes take place in a haphazard way, or whether they are governed by rules essentially the same in all the different groups. Zoologists do not, on the whole, appear to have recognized that they are confronted here with a fundamental problem. The great variety of existing pelagic larval forms, often within uniform and well defined groups, suggests that no general rules are involved. The literature, at any rate, demonstrates considerable uncertainty and confusion. The great increase in our factual knowledge of development has not led to a corresponding increase in our understanding of the interrelationships of the facts, still less to a comprehensive phylogenetic theory in terms of which the facts might be satisfactorily interpreted. Thus there are still different opinions about such an important basic question, as to whether direct or indirect development should be considered the original type.

In the course of elaborating my *Bilaterogastraea* theory (1955), there occurred to me, so to speak as a by-product, an idea about the nature of the original metazoan life cycle and the evolution of pelagic larval forms that seemed worth pursuing. A tentative application of the idea to some of the larger groups gave such positive and encouraging results that I decided to interrupt for the time being my studies of "genealogical trees" and to devote myself instead to a detailed investigation of the life cycles. The outcome was a general and, I believe, well-founded theory that is applicable to all the Metazoa.

It is obvious that an investigation of this kind requires a broad base, not only of morphological and ontogenetic, but also of ecological facts. In the course of time a large fund of knowledge has been amassed. Nevertheless in many cases I have been unable to find the information I required, and have thus been compelled to make my own primary studies. Brief accounts of these studies have in several cases been included in the text.

From what has been said above it will be clear that this work is not intended to be a text-book in the current sense of that word. On the other hand, I have aimed at a kind of presentation that will, I hope, make the contents accessible to all students with a basic academic knowledge in zoology. The contents have for many years formed an essential part of my lectures at the university, principally in advanced studies, though because of the fundamental nature of the subject, aspects of it cannot be entirely disregarded even in elementary teaching.

Some will perhaps find my theory daring, even provocative, and there is no doubt that many of my results are rather unexpected. When confronted with phylogenetically interesting questions on which the theory might shed new light, I have not, generally speaking, hesitated to draw whatever conclusions were demanded. If no attempt is made to solve or at least approach the problems, understanding will certainly not deepen.

Hardly any new biological theory has been put forward without generating opposition, at least in the beginning. Some find it difficult to adopt lines of thought that are more or less opposite to the ones they are accustomed to. This remark was made to me by a well-known zoologist after the exposition of my *Bilaterogastraea* theory. Probably I shall meet with criticism of this kind again now, and perhaps there will also be objections to my interpretation of certain special points. I am interested to hear differing views on any of the questions I have raised.

I would like to add that this work could hardly have been carried out without a great deal of travelling, in some cases to very distant countries. For this purpose financial aid was obtained from the University of Uppsala, the Government Scientific Research Council, the Nordic College for Marine Biology, the International Indian Ocean Expedition, and the American Museum of Natural History. My visits to the Lerner Laboratory on Bimini (Bahamas) have been of particular importance. I wish to offer my thanks to Mr. Michael Lerner, Miami; the Director of the American Museum of Natural History, Dr. James A. Oliver; the Assistant Director, Dr. Joseph M. Chamberlain; the Resident Director of the Bimini Laboratory, Mr. Robert F. Mathewson; and many others who in different ways have facilitated my investigations. My particular thanks are due to my friend, lecturer Bror Forsman, with whom I have had valuable discussions during many summers spent with him at Klubbans Biological Station; to lecturer Lennart Gidholm for the critical perusal of certain parts of the manuscript; to lecturer Birger Pejler for reading the section on rotifers; to Mrs. Kerstin Ahlfors for the fine execution of the drawings; and finally to my wife who has been of great assistance on my journeys, especially in the acquisition of material.

This work is basically a direct translation of my book in Swedish of 1968.

December, 1971   *Gösta Jägersten*

Contents

General Section

Introduction

Ever since Haeckel established his famous theory of recapitulation (biogenetic "law") juvenile forms, the larvae especially, have played a prominent rôle in the discussion of phylogenetic problems. In some cases, e.g. the correct taxonomic placing of the Rhizocephala and certain parasitic gastropods (*Entoconcha*, etc.), they have been of quite decisive importance.

According to extreme recapitulationists, still greater importance must, however, be ascribed to the larvae. With certain exceptions they should supply a direct picture of the ancestral forms in the adult stage ("palingenetic" larvae). It is hardly likely that this extreme application of the theory of recapitulation still has any adherents today. It is now quite generally believed that recent larvae only repeat the conditions of the larvae of the ancestral forms (this conception has an energetic advocate in de Beer, 1958). The pelagic marine larvae should, even if some of them are very ancient, represent the result of the adaptation of juvenile forms to life in the open water, and this adaptation should have taken place in different instances. Thus direct development should be the original process (an idea not shared, however, by all authors). Certain zoologists have even gone so far as to suggest that different forms of such morphologically uniform types as, for instance, the *trochophora*-larva were the result of convergent evolution. In other words, according to this idea, the significance of the larvae from the point of view of phylogeny is nil, at least as far as the understanding of the phylogeny of higher taxa is concerned.

It appears to me quite certain that the extreme recapitulationists have overestimated the importance of the larvae in their attempts to visualize the adult stages of the ancestral forms, but it seems equally certain that their opponents are guilty of underestimation. This work tries to put these things right, being an attempt to elaborate a general view of the larvae and the life cycles of the Metazoa with special attention to the more primitive phyla and with consideration of function, biotope and ecology.

I wish to assert that upon this basis new ideas about the adults of the ancestral forms and thus about the phylogeny of the Metazoa in general can be obtained. It is also obvious that we cannot arrive at a clear and universally applicable conception of the fundamental problem of re-capitulation so long as we hold an incorrect or hazy idea of the larvae and ontogeny from the phylogenetic point of view. A study of the literature on this extensively treated problem makes this quite clear.

By the above I want to underline the importance of this subject to taxonomy and thus to all branches of natural science based upon it. Up to now taxonomy has been based too much upon a single stage in the life cycle—the adult stage. It is necessary, particularly concerning lower systematic units, to pay more attention to the *entire* life cycle and to the relationships between its different stages. This applies especially to forms with larvae and adults in separate biotopes.

It is often asserted or taken more or less for granted that the larvae are adaptive forms and for this reason of minor importance in taxonomic and phylogenetic investigations. I wish to stress the danger involved by a general and uncritical application of such a view. There are, in fact, many examples of adults with an adaptive faculty that is certainly no less than that of the larvae. Furthermore, the phylogenetic importance of the characters is not dependent on whether or not they are adaptations, but on the place of their appearance in the evolutionary tree.

In my papers about the *Bilaterogastraea* theory (1955, 1959) I have up-held the view that the occurrence of pelagic juvenile forms is the original situation in all Metazoa. In other words, when the radially symmetrical *Blastaea* forms became bottom-dwellers (and as a result became bilaterally symmetrical) the juveniles remained in the pelagic zone, where they con-tinued to swim about with the aid of their ciliary cover. This evolution thus implied that the life cycle was split into two phases, viz. a pelagic juvenile and a benthic adult phase ("pelago-benthic life cycle"). At the transition to the benthic region that took place in every individual ontogeny, there occurred certain changes which made the animal fit for the radically different new biotope. These changes, which in *Bilatero-blastaea* and the subsequent early metazoans were relatively simple, con-stituted the initiation of metamorphosis. The continued evolution of both the pelagic young and the benthic adult, which included a more intimate adaption to their respective biotopes, implied that this metamorphosis became more extreme in the majority of the evolutionary lines leading to the recent groups. The two developmental stages diverged to a varying extent, and the juvenile form became a "larva".

The theory presented in this book is not an unfounded speculation. Admittedly, it was originally presented as, so to speak, a by-product of the *Bilaterogastraea* theory (Jägersten, 1959, p. 87), but it is nevertheless based directly upon actual facts and conditions at the present time. If we examine the situation within the different large groups, we find that the constellation pelagic larva and benthic adult occurs in the majority of them (see p. 213).

Already this wide distribution (see also Fig. 57) leads us to surmise that the pelago-benthic life cycle is so ancient that it dates back to the common ancestral forms of all metazoans. There is nothing to support the assumption that a holobenthic life with direct development should be original in the recent phyla, and that pelagic larvae should have evolved later on in the special lines leading to the different groups. Our accumulated experience of ontogeny, based upon numerous investigations, tells us that on the contrary the existing life cycles with direct development are secondary phenomena that have evolved independently in several different instances, and are generally connected with changed environmental conditions and/or an increased quantity of yolk in the egg.

We have irrefutable proof in support of the pelagic larval life as an original feature. In cases of benthic embryonic development inside the egg membrane there are stages which more or less clearly resemble the pelagic larva (*trochophora*, *veliger*, *pilidium*, etc.) in related forms and groups. Such embryonic stages thus exhibit remains of pelagic adaptations —characters that are no longer of any use, but which are stubbornly retained.

This is so well documented that it would be superfluous to enter into the matter more closely, were it not for its great importance and the fact that there exist authors who still assert that direct development is the original condition.

Before embarking on a treatment of the different groups I must introduce certain terms and explain their meaning. I shall also propose theses which will be given more detailed discussion in the treatment of the different groups and in the general section of the book.

Primary larvae

In speaking here about larvae, I refer to a particular kind which for greater clarity will be called "primary larvae", in order to indicate that they are derived back directly from the larvae of the first Metazoa.* It has already been mentioned that the first divergence between the juvenile phase and the adult phase in ontogeny took place, when the holopelagic *Blastaea*, the ancestral form common to all Metazoa, took to life at the bottom in its adult phase. The qualification "primary" thus neither refers to any special characters in the morphology of the larvae—which in any case may become very different in the various evolutionary lines leading to the recent groups—nor to the degree of complication of structure, but to the fact that the pelagic larval type has persisted in ontogeny without interruption since its first appearance.

Examples of primary larvae are such well-known forms as *cyphonautes*, *trochophora*, *actinotrocha*, *tornaria*, *auricularia*, etc. It must be pointed out that in certain cases where two or more larval forms appear in one and the same ontogenetic sequence, e.g. *trochophora* and *veliger* in certain gastropods, they should still be considered as primary larvae (the *veliger* larva is merely a more advanced stage and cannot be termed "secondary larva", as is explained below).

When dealing with the primary larvae we must stress the principle of the *divergence between two phases of the life cycle—the pelagic larval phase and the benthic adult phase—the two phases being separated by metamorphosis*. The essential cause of this divergence is, as I have shown, the adaptation of these phases to their respective biotope (here I make allowance for the possibility that some characters are only indirectly conditioned by adaptation). In asserting that the pelagic larvae show adaptation to the environment I am opposing the extreme view that these larvae should always recapitulate ancient adults. Although the "principle of divergence" has led me to to an idea identical with that presumably held by most zoologists today, I must, nevertheless disagree strongly with the

---

* Unfortunately I find it necessary to use the terms "primary larva" and "secondary larva" although they have been used in various senses by several earlier authors, the distinction coinciding in part with that adopted by me. This applies, for example, to Schmidt (1966). Since the principles I have proposed have not been vindicated before, nobody has previously been able to apply consistently the division based upon them.

prevalent opinion that the pelagic larvae are not important in phylogenetic considerations (for more detail, see below).

One consequence of the principle of divergence that is worth stressing is that the fully developed pelagic larva must to some degree be considered a final stage. Embryogenesis gives rise to an individual living a life that is often essentially different from that of the adult, with different loco-motion, feeding, etc. The larva very often possesses special highly de-veloped organs, and its tissues are to a great extent well differentiated and without embryonic features. Examples are the *pelagosphaera* of the sipunculids, the *cyphonautes* of the bryozoans, and larvae of the ento-procts. It is obvious that this phenomenon is connected with the fact that many organs and parts of the body do not develop any further, but are discarded or otherwise lost at metamorphosis. Features of certain ento-procts are worthy of particular mention. Here the juveniles develop from buds inside the larva, which does not undergo metamorphosis but dies immediately after the "birth" of the juveniles. The organization of these entoproct larvae is "finished" to a high degree, and the majority of their tissues exhibit no embryonic features (Jägersten, 1964).

Adultation

It is a well-known fact that the marine pelagic larvae of benthic adults very often exhibit both larval and adult characters, existing side by side. This can be exemplified by again referring to the *veliger* of the molluscs. In this larva the velum is an exclusively larval character, while shell and foot are adult features.

The majority of the larval characters are more or less obvious adapta-tions to the pelagic life and are therefore lost at metamorphosis. The adult characters are of course retained and develop further during and after metamorphosis.

In maintaining above that the divergence between pelagic larva and benthic adult is the result of the adaptation of these two stages to their respective biotopes, I have deliberately disregarded the possible occur-rence of adult characters in the larvae, since I was then faced with the problem of explaining the differences. I must now explain the similarities between the two stages, which are often found and which are a result of the occurrence of adult characters in the larvae.

Partly because I have arrived at the conclusion that a pelagic larval life is original for all Metazoa, and partly because of the existence of special ontogenetic features in several groups (see p. 216), I cannot accept an explanation based on the assumption that juveniles with certain adult characters have, in different evolutionary lines, ascended into the pelagic zone and there become larvae. On the contrary adult characters (the phylogenetic appearance of which always occurs in the adult phase, see p. 221) must have been shifted to the pelagic (larval) phase. In other words this phenomenon which I shall call "adultation" is essentially a special case of what for a long time has been known as "acceleration", a phenomenon that so far seems to have been applied only to life cycles with direct development. The unknown forces that lie behind adultation will be termed as "adult pressure". (For a further elucidation of adultation see p. 218.)

A shifting of the initial development of the adult characters to the pelagic larva is in reality a very common phenomenon which I shall deal with in connection with the different groups. In the case of larvae that differ greatly from the adult the phenomenon is of particular importance, since the development of adult characters during the pelagic phase greatly facilitates metamorphosis.

Sometimes relatively simple characters appear to be developed unnecessarily early. A case of this is, for example, *Protodrilus*, where in certain species the adhesive lobes of the adult are already developed in the earliest *trochophora* stage in spite of the fact that they are only functional in the creeping worm. In other members of the same genus they are, however, developed at a much later date (Jägersten, 1952). Another example of a very early development is met with in many of the mollusc larvae where the shell or at least the "shell-gland" can already be distinguished in an early *gastrula* stage.

In both these examples and in numerous other cases there can be no doubt about the interpretation. The phenomenon is in reality so common that *we can designate the phenomenon of adultation as a general principle among the metazoans* (see p. 217 et seq.).

We can now progress further. There is no reason to assume that adult pressure has had an effect only on recent adult characters. If such characters could be shifted to the pelagic larva, the same conditions must also have prevailed during still earlier phases of evolution. In other words, there can hardly be any doubt that characters distinguishing the adults of the ancestral forms were shifted (accelerated) to their larvae. Since the larva is known to be often more conservative than the adult, we have to reckon with the possibility that some of the ancient adult characters which had been shifted to the larva can be retained by it even after they have disappeared in the adult. *A priori* it can thus be imagined that they also occur in recent times. In fact a closer examination of the larvae of the different groups of Metazoa shows that side by side with typical characters of the primary larvae (pelagic characters) and features of the recent adult a third type, distinct from the other two, frequently occurs. Such characters must be given close attention. They may be *ancient* adult characters which can supply valuable information about the nature of the ancestral forms in the adult stage, and thus may open up new possibilities for the elucidation of the phylogeny of the Metazoa.

I will give an illustrative example. It is well known that certain holothurians harbour some strongly transformed worm-like parasites that can be classified as gastropods solely on the basis of their larvae (*Entoconcha*, *Enteroxenos*). No zoologist today is likely to assert that the adults of the ancestral forms of these parasites were without a shell, etc., and that their entire aspect was unlike typical recent gastropods. As has been pointed out already the shell is a recent adult character in the vast majority of veliger larvae. (In the larvae of the parasites mentioned it is, however, an ancient adult character.) Let us now assume for the sake of argument that with the exception of *Entoconcha* and its closest relatives all gastropods had become extinct without leaving any fossil documentation. The shell in the larvae of the parasites in question, being neither the result of an adaptation to the pelagic life nor a recent adult character, would nevertheless indicate something of which we now have a more direct knowledge—that the adults of the ancestral forms were provided with a dorsal shell.

Direct development

From what has been said in the foregoing about the primary larvae it follows that direct development, viz. more or less far-reaching adultation, including loss of the pelagic larval characters, whether it occurs in all species within a group or only in some isolated forms, is in the end derived from a pelago-benthic life cycle with a primary larva. Direct development has arisen independently many times within the evolutionary tree of the Metazoa. Thus it is no surprise to find that it exhibits a considerable number of variations; the change in the original life cycle has progressed to a varying degree in different cases. It is accordingly only to be expected that zoologists are frequently at variance as to whether a certain case represents direct development or not.

It can be seen that it is necessary to distinguish between two main types of direct development which I call "primary" and "secondary".

Primary direct development has evolved immediately from a pelago-benthic life cycle with primary larva. Secondary direct development has on the other hand always appeared only after a period of evolution with a new larva (a "secondary larva", see below). In both types the entire life cycle is as a rule limited to the same main biotope, either the benthic or the pelagic zone.

Secondary larvae

It is evident that in the life cycle with primary direct development new larvae, "secondary larvae", can be differentiated. In such cases a new divergence between the juvenile form and the adult has arisen. As a rule one or the other of these stages, or more rarely both, have become subjected to frequently very profound changes involving a new metamorphosis. Exceptionally secondary larva and adult live in the same biotope, but as a rule the biotopes are radically different.

After this account of the different major types of life cycle in the Metazoa, the technical terms can be used without further explanation. The phylogenetic connections are summed up in the following scheme:

Ontogeny with primary larva and metamorphosis
(primary pelago-benthic life cycle)
      ↓
Direct development (primary)
      ↓
Ontogeny with secondary larva and secondary metamorphosis
      ↓
New direct development (secondary)

I wish to stress that in the above scheme I do not insist upon the existence of any sharp differences between the types distinguished. On the contrary there exist numbers of intermediate cases. Most ontogenies are, however, relatively easily classified. The greatest importance of the types distinguished is, however, that they represent steps in a certain evolutionary direction. In the different species or groups evolution has just reached one or another type in the established scheme or occupies a place somewhere between two of them.

I shall conclude this introduction with two theses that in my opinion are almost truisms, but which may nevertheless facilitate understanding of subsequent parts of the text. They are:

1. Larval characters—at least those that are adaptations to pelagic life— have always made their first phylogenetic appearance in the pelagic phase of the life cycle.
2. Adult characters—at least those that are adaptations to benthic life— have always made their first phylogenetic appearance in the benthic phase of the life cycle (for a more detailed treatment see p. 22).

Special Section

Spongiaria

In the sponges, where the adults are always benthic and sessile, there occurs a larva with cilia covering a part or the whole of the body. After a brief period of swimming the larva, which as a rule leaves the mother animal in a fully developed condition,* becomes a sedentary and undergoes metamorphosis leading to the adult form. We thus have reason to suppose that in the common ancestral form the life cycle was similarly split into a pelagic larval and a benthic adult phase. The larva must without doubt be interpreted as a primary larva. It is always lecithotrophic—a characteristic of course connected with its lack of an enteron.

Now, is this lack to be considered a primary or a secondary feature? Although according to the *Bilaterogastraea* theory the free-living adult of the ancestral form possessed an alimentary cavity in the shape of the invaginated ventral side (see Jägersten, 1955, Fig. 1c), I am inclined to think that in the evolution of the sponges such a cavity has never existed in the *pelagic* phase of the life cycle. This opinion is based on the following reasoning:

Once the common ancestral form of all Metazoa, including the sponges, had descended to the bottom according to the theory, no enteron existed either in the juvenile form (the larva) which continued life in the pelagic zone or in the adult which then and subsequently led a benthic life. When in the adult a simple enteron was formed, i.e. when the *Bilaterogastraea* evolved (see Jägersten, 1955, p. 328), this transformation must be considered as an adaptation to the benthic biotope, and for this reason it seems unlikely that at this time the pelagic larva had passed beyond a simple blastula stage. The metamorphosis that took place in every generation at the transition to benthic life essentially consisted of the invagination of the ventral side and resulted in the primitive enteron in

---

* According to Trégouboff and Rose ("Manuel de planctologie Méditerranéenne", 1957, p. 259) *Cliona* is oviparous.

question. Strictly speaking, the digestive tract of the Metazoa should thus be an adult character (see also p. 224).

It is probable that these conditions still occurred at the beginning of the special line of evolution that led to the Spongiaria (see ancestral tree, p. 214). Here, however, the so far free-living adult became sessile and great morphological changes resulted. The pelagic larva on the other hand remained largely unaltered. In the majority of forms the blastocoele was, however, filled with mesenchyme.

The above discussion implies that the lack of enteron (and the occurrence of lecithotrophy) in the larvae of the Spongiaria is in all probability an original feature. In that respect these larvae are unlike all other metazoan larvae devoid of enteron. This would agree well with the fact that in sponges the process of invagination, which is usually interpreted as gastrulation, takes place only after the individual has become sedentary.

Because of the divergence between larva and adult, which became very marked after adoption of sedentary life by the latter, the larva after settling down had to undergo fast and far-reaching metamorphosis. In contrast to many higher groups the divergence has not been moderated by any distinct adultation of the pelagic larva. It is, however, to be noted that some larvae possess internal skeletal needles. These structures are of course an adult feature.

It is of considerable interest that the sponges have a pelago-benthic life cycle, and that this contains a larva, albeit one of simple organization. This resemblance to other metazoans can be adduced as an argument in the discussion about the phylogeny of the sponges. However, I shall not elaborate on this question here. I only want to point out that I still consider it more probable that the sponges share a common ancestor with the other groups of multicellular animals than that they have evolved independently of them (the Parazoa hypothesis).* I should mention that in my previous discussion of the sponges (Jägersten, 1955), I only wanted to show that the group *can* without difficulty be supposed to have originated from a *Bilaterogastraea*. In the organization of the sponges nothing can be established that would point directly to their descent from bilateral forms, but it would surely be most remarkable if the feeble bilateral

* The Parazoa hypothesis which is based mainly upon the similarity of the choanocytes of sponges to choanoflagellates has been criticized already by Metschnikoff (1886, p. 133). Recently its credibility has been reduced by a discovery of Nørrevang (1964), who with the aid of electron microscopy established that choanocytes are also found in the epidermis of enteropneusts (*Harrimannia*).

symmetry found in the metazoans at the time of the splitting-off of the branch of the sponges (see the ancestral tree, p. 214) had not been obliterated in the course of the immensely long time during which this group has been sessile.

Cnidaria

Within all three groups of the Cnidaria the majority of forms possess a pelago-benthic life cycle which, however, in Hydrozoa and Scyphozoa is complicated by the occurrence of metagenesis. This complication by no means precludes the application of the theory of the originality of the pelago-benthic life cycle to the Cnidaria. As far as can be judged, metagenesis is a secondary phenomenon which arose after the differentiation of the Cnidaria (see Jägersten, 1955). The theory in question leads to the conclusion that the holopelagic life cycle in *Siphonophora* is also secondary.

Irrespective of metagenesis the cnidarian life cycle generally takes the following course. The sexual products are freely discharged into the water where fertilization takes place. The egg, which is in suspension, gives rise to the pelagic *planula*-larva which is as a rule covered with cilia all over the body (see p. 2). In certain cases even the blastula is ciliated and free-swimming. In all forms which are not holopelagic the larva attaches itself with the anterior end (or exceptionally sideways). It then develops tentacles around the mouth and gradually becomes a sexually mature polyp.

There are, however, certain divergences from this general course of development, for example in the case of brooding. This phenomenon is rather variable, but the egg is always retained for some time upon or within the body of the mother animal where development takes place for varying lengths of time.

In the Hydrozoa with medusae the eggs are as a rule discharged in an unfertilized condition. In forms with sessile gonophores, on the other hand, internal brooding generally takes place, and the young leave as fully developed *planula*-larvae. Accordingly the gonophores generally form eggs that are larger and contain more yolk than those of the free medusae.

The genus *Tubularia* and its relatives are examples of forms which exhibit a considerable prolongation of the development inside the mother animal, with the result that an almost fully developed polyp, the tentacu-late so-called *actinula*-larva, leaves the gonophore. Thus no ordinary free-swimming *planula* occurs (this is incorporated as a stage in embryo-genesis). The *actinula*-larva differs from the polyp of *Tubularia* principally by having an incomplete set of tentacles and, of course, by its lack of a stalk, but the ensuing changes are gradual. Thus we have here a type of direct development. In *Tubularia dumortieri* development is closer to the original conditions, the *actinula* being preceded by a pelagic *planula*.

As is well known direct development occurs also in *Hydra*. Here the young embryo, while still in the ovary of the mother animal, surrounds itself with a chitin-like shell within which embryogenesis continues to take place. After a period of rest it liberates a young animal with small incipient tentacles, at least in certain forms. It is possible that here as in so many other cases the transformation to direct development is connected with immigration into fresh water.

Examples within *Corymorpha* show that direct development may also take place entirely outside the mother animal. The egg, enclosed in a covering, is deposited upon the bottom. At hatching an *actinula*-like polyp is freed and this attaches itself by the aboral pole.

In Anthozoa, too, brooding with accompanying elimination of the pelagic *planula*-larva occurs. According to information given by Pax (Anthozoa, Tierwelt der Nord- und Ostsee) the young of the genera *Isozoanthus*, *Parazoanthus*, and *Epizoanthus* (all belonging to the *Zoantharia*), which live in the North Sea, do not leave the mother animal until they have developed a cuticle. As an example of a holobenthic life cycle in Actiniaria the species *Halcampa duodecimcirrata* can be mentioned (Nyholm, 1949). Here the larva has lost the ciliation of the body surface and is consequently no longer able to swim.

Another divergence from the course of development typical of cnidarians consists of a prolongation of the pelagic phase beyond the *planula* stage. Examples of this process are found especially within the Ceriantharia. In *Cerianthus lloydii* the planktonic life last from three to four months (Nyholm, 1943), and during this time the two circles of tentacles characteristic of the adult are developed. The transformation from a *planula* devoid of tentacles to this *synarachnactis* stage, in which it is ready to adopt life on the bottom, is gradual. The transformation process may probably be fastest just at the time of the change of biotope.

A still more extreme extension of the pelagic phase has taken place in *Arachnactis albida*, where larger individuals—the body, including the tentacles, may reach a length of at least 5 cm—may even start to develop gonads. Since a benthic form is unknown, the possibility is considered that in this case no change to benthic life ever occurs. If this is so, then the species supplies an example of the genesis of a holopelagic life cycle on a neotenic basis (see also p. 234 and Fig. 58).

Exceptionally the Actiniaria, too, may show the development of a crown of tentacles during pelagic life. This is maintained by Wietrzykowski (1914) for *Edwardsia beautempsi*.

Another and more common adult character in the pelagic phase of the actinians and other groups are septa. It could also be mentioned that in the larvae of certain actinians not only well-developed ovaries but also embryos in the gastrula stage are encountered (Carlgren, 1924). This may be a case of dissogony. The appearance of these different adult characters in the pelagic phase of the life cycle ("adultation" of the larva) agrees well with the situation in many forms belonging to other groups (see p. 217 *et seq*.).*

Of the two main types of divergent ontogeny in Cnidaria distinguished above we must first consider the type in which no pelagic ciliated *planula*-larva develops, but where instead there is direct development within the mother animal or in its vicinity, or at least in the same biotope. This type of development is found in only a small section of the cnidarians. As far as I know, it has not been established at all in systematic units as large as Octocorallia, Ceriantharia, and Madreporaria. In fact, few cases are known beyond those mentioned above. It is very interesting to find that forms with direct development of this type generally possess relatives with a typical pelagic *planula*. In other words, a pelagic *planula* is characteristic not only of the three main groups of the cnidarians but also of the majority of the lower-ranking systematic units.†

There is thus no reason to doubt that the pelagic *planula* was originally characteristic of all cnidarians, and that the benthic direct development is

---

* It is hardly necessary to point out that structures like the gonads, although without doubt adult characters, are not to be interpreted as an adaptation to benthic life. It is to observe that not all adult characters represent such adaptations.

† The Stauromedusae possibly supply an exception. In *Haliclystus* the larva is said to be devoid of cilia and not to leave the bottom. But the conditions within the group are not very well known, and it is quite possible that a typical *planula* also occurs within this group.

It might also be pointed out that the ontogeny of the Antipatharia is still unknown.

a later acquisition that has arisen independently at different points of the ancestral tree of the group. The common ancestral form of the cnidarians was in all probability benthic in its adult stage and possessed a pelagic primary larva.

Because of the simple organization in cnidarians at all stages the transformation of the *planula*-larva into the benthic adult does not imply a very radical metamorphosis. For this reason Geigy and Portmann (1941) prefer not to speak of larva and metamorphosis but, instead, of "primäre direkte Entwicklung", especially since in their opinion "kein einziges Merkmal, das als transitorische, also larvale Bildung angesprochen werden könnte", is found in the *planula*-larva. Quite apart from the fact that in this paper I use the term "primary direct development" with a different meaning (see p. 8) I must repudiate calling ontogeny via a *planula* a direct development. The transformation is admittedly not very radical, but the *planula* is nevertheless not entirely without purely larval characters. In addition to the ciliation of the ectoderm, which is by no means always present in the adult, the anterior end of the larva in many cases, perhaps always, has sensory cells. These are often concentrated into a more or less pronounced sensory organ. Below it we find a thickened layer of nerves (see e.g. Jägersten, 1959, Fig. 3). All this is lost at metamorphosis. It is furthermore inadmissible to disregard the general dissimilarity between larva and adult, and the speed of the transformation which as a rule starts immediately after the adoption of a benthic life. If the development of the *planula*-larva should be termed direct, then for the sake of uniformity the same must apply, for example, to numerous *trochophora*-larvae in *Polychaeta*, and this would lead to unnecessary confusion. I must also draw attention to the essentially modified ontogeny in, for example, *Hydra*, which must be termed direct, and according to my terminology is a "primary direct development". The same designation ought also to be applied to the above-mentioned internal development in certain zoantharians, and perhaps also to the development characteristic of ceriantharians with prolonged pelagic life such as *Cerianthus lloydii*. It is perfectly understandable that in cases of altered ontogeny there will be some doubt about classification (see p. 8).

Where, on the other hand, typical *planula*-larvae are present, no difficulties occur. According to the terminology used here they are without doubt "primary larvae", even if in comparison with those of the Coelomata they are very simply organized.

It would be quite in accordance with my general views, if the pelagic larvae revealed traces of a bilaterally symmetrical organization even in other adult features than those connected with septa and stomodaeum. It is therefore interesting to find that within the Zoantharia the so-called *zoanthella*-larva provides an example of this (Conklin, 1908). Here the bilaterality finds its strongest expression in the ventrally situated, longitudinal band of cilia (Figs. 1C and D). Unfortunately the early development of this larva is entirely unknown. Its organization is, however, such that it seems unlikely that the band has arisen in a pelagic larva. I find it

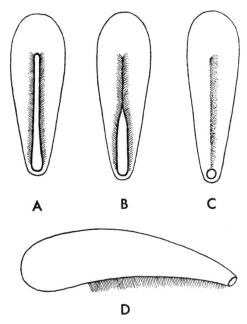

FIG. 1. Diagrams illustrating the supposed course of evolution leading to the *zoanthella*-larva. A, original stage with long oral fissure bearing a ciliated band on either side. B, stage with anterior part of oral fissure closed. C and D, present condition. A–C, ventral view of the body, D, body seen from the left.

considerably more credible that we are concerned here with reminiscences from a time when the adult was a free-living bilateral organism, moving on the substratum by means of ciliated bands, situated one on either side of the strongly elongated mouth. Before the adult became sessile, adult pressure had already resulted in the shifting of the development of these ciliated bands to the pelagic larva (Fig. 1A). After the establishment of the sessile mode of life they were, however, lost in the adult. It may thus be imagined that we are concerned here with an ancient adult character (see p. 7).

As so often happens with ancient adult characters in larvae the ciliated bands became modified (for such modifications see p. 223). It seems as if this modification was related to a shortening of the originally long oral fissure, effected by coalescence of the edges in an oro-caudal direction (Fig. 1B). Thereby the two ciliated bands were gradually fused into a single one in front of the mouth opening, the process ultimately resulting in the present form (Fig. 1C and D). If this interpretation is correct, then the change is in close agreement with that in certain Deuterostomia, e.g. the *tornaria* of the Enteropneusta (cf. p. 245, and Fig. 59B).

Two observations are worth making here. Firstly, in one *zoanthella*-larva a pore has been recorded in front of the ciliated band; this might be explained by the assumption that no coalescence had taken place at the extreme anterior end. Secondly, in a number of other specimens it could be observed that the band was double (see Conklin, 1908, p. 183). The latter condition strengthens the interpretation given here, that an originally long oral fissure has been partly obliterated (cf. Fig. 1A).

It is not easy to tell why only the *zoanthella*-larva should exhibit the ancient adult character which the ciliated band seems to represent. This might be due to the fact that in the other cnidarians the development of the band has not been shifted ("accelerated") to the pelagic phase of the life cycle. It appears, however, more likely that some other larvae, too, have undergone the same adultation, but that the band has been retained only in the *zoanthella*. The reason for the retention here may be that, as a result of the pronounced lengthening and mutual cementation of the cilia, the band became the effective swimming organ (in the form of an undulating membrane) which it still is in this larva. It is worthy of notice that the general ciliation of the body consists of extremely small cilia, which in comparison with the powerful band can hardly be of any importance for locomotion (my own observation on material from the Florida current).

Another well established case of bilateral symmetry in the *planula*-larva occurs in *Funiculina*, where according to Berg (1941) the stomodaeum is formed somewhat towards the ventral side. It is possible that this is the first indication of the bilateral symmetry of the colony found throughout the entire group Pennatularia, but also in this case an explanation is necessary. In my opinion this explanation is given by the

Bilaterogastraea theory, in that the bilateral symmetry of the colonies may be derived from that of the solitary ancestor.

The absence of bilateral features in the great majority of the *planula*-larvae is evidently no argument against the *Bilaterogastraea* theory. Because of the adaptation of the *planula* and the preceding embryonic stages to pelagic life ontogeny may well also have become modified towards radial symmetry, as I have already pointed out in an earlier paper (Jägersten, 1959, p. 87, and Fig. 2). We must also allow for that kind of adult pressure which arose *after* the adult had become radially organized and which must act in the same direction as the pelagic biotope. This adult pressure might even have played the greater rôle.*

Among the cnidarians it is known that in many cases fertilization takes place outside the body of the mother animal, and that the uncloven egg and all later embryonic stages remain suspended in the water. It is fairly common to find cilia already developed in the blastula-stage (cf. p. 242). There is no doubt that external fertilization is the original condition and this is fully in agreement with the *Bilaterogastraea* theory (Franzén, 1956; Jägersten, 1959).

The egg often contains a considerable amount of yolk. Connected with this is that the pelagic larva is lecithotrophic and usually has no mouth. There are examples where the mouth is absent even in larvae of forms where the original type of gastrulation, i.e. invagination (Jägersten, 1955, 1959) is retained (*Aurelia*, *Cyanea*, and other scyphozoans). Among Hydrozoa this type of gastrulation is unknown; it has been replaced by the immigration of isolated cells (polar or multipolar immigration) or by delamination.

It must be stated that lecithotrophy is dominant among the Cnidaria; within Hydrozoa, Scyphozoa, Octocorallia, and Madreporaria no exceptions seem to be known. Planktotrophic larvae occur within Ceriantharia and Actiniaria, but not in all forms of these subgroups.

In spite of the fact that from the purely statistical point of view lecithotrophy predominates within the cnidarians, the feature can nevertheless not be considered as original for the group. As far as we can judge

---

* With the experience now at my disposal I no longer consider it necessary, as I stated earlier (Jägersten, 1959, Fig. 2), that the transformation of the larva to an external radial symmetry must have taken place simultaneously with the transformation of the adult. The change of the larva might well have taken place considerably later than that of the adult.

planktotrophy must be the original condition (see p. 224). It should be pointed out that this active way of feeding occurs in such groups as Cerianthariae and Actiniaria—groups which are primitive also in other respects (absence of formation of colonies and lack of skeleton). Of greater importance however, is, the obvious connection existing in the cnidarians between lecithotrophy and altered gastrulation. Wherever gastrulation is effected by delamination or by immigration of cells, either polar or multipolar, we also find lecithotrophy of the larva and never planktotrophy. Wherever an active mode of feeding has been established we find the original type of gastrulation, i.e. invagination. Lecithotrophy may occur in larvae where embryogenesis includes invagination, but in this case there are certain deviations. An example of this is supplied by *Pachycerianthus*, which, in spite of the great mass of yolk in the egg, nevertheless exhibits invagination, a stubborn retention of an original feature. The divergence from the typical consists mainly of the detachment of a large lump of yolk from the entoderm with the result that the yolk finds its way into the gastrocoel (see Nyholm, 1943, Fig. 19).

Pachycerianthus exhibits conditions which very strongly support my opinion, based on other arguments (Jägersten, 1955, 1959), that the original method of gastrulation is invagination. Here invagination takes place in spite of the great quantity of yolk. It might have been expected, as pointed out by Nyholm (1943, p. 206), that delamination would prove a better solution with eggs of this type. But invagination nevertheless persists.

In establishing the connection between feeding and type of gastrulation I have of course not expressed any opinion about the causal connections. Altered gastrulation may appear to cause lecithotrophy, but conditions are hardly as simple as that. Instead, each phenomenon may result from the rich supply of yolk in the egg usual in cases of this kind, but this explanation is also open to objections. It is safest to restrict the discussion to stressing the fact that a certain mass of yolk in the egg is a necessary condition for the evolution of lecithotrophy. The larva must dispose of a quantity of yolk sufficient for the maintenance of life during the pelagic phase and the transformation into the usually benthic adult (see also p. 240 *et seq.*).

It is, however, evident that in the Cnidaria there is an interrelation not only between a more or less ample supply of yolk and altered gastrulation, but also between these phenomena and lecithotrophy. As has already been established it is the connection between altered gastrulation and lecithotrophy that is most important. This must indicate that in the cnidarians a planktotrophic larval life is the original situation and that lecithotrophy is a secondary phenomenon which must certainly have arisen independently on different occasions within the group. (For discussion of the change to lecithotrophy see also p. 225.)

What has been said above about the yolk in no way refers to the absolute amount of yolk necessary for the changes. In this respect there are of course great differences. If the length of the pelagic life of the larva is short, as is usually the case in Hydrozoa, the change to lecithotrophy is possible even without a major increase in the quantity of yolk.

The absence of typical invagination even in such hydrozoans that have eggs which are relatively poor in yolk might possibly be explained by the shortness of the pelagic life of the larva together with the fact that its transformation into a polyp with active feeding is a comparatively simple process which requires very little material. In other words, here the conditions for a change of gastrulation (including elimination of the mouth) have existed without the storage of a greater mass of yolk, and the change has then been effected.

Because our knowledge about the development of the cnidarians is still incomplete, it is hardly possible to carry reasoning any further. Renewed investigations with a wider scope and with attention to these problems would therefore be welcome.

Already, however, we can discern in the ontogeny and mode of life of the larva in the three groups of the Cnidaria the same sequence to which I have been led earlier by the *Bilaterogastraea* theory. The most primitive features are found within the Anthozoa, the most altered within the Hydrozoa.* I must, however, underline that the changes mentioned have probably not taken place along a single evolutionary line but along several parallel lines.

Indirectly it can be seen from the foregoing that when using the name *planula* for the larval form which I have found to be a constituent of the original cnidarian life cycle, I mean a pelagic form covered with cilia over the whole or the greater part of the body. I have thus not referred to the *planula* as a form which should be devoid of mouth opening and entoderm cavity, neither have I made any distinction between forms with

---

* On the basis of comparative examination Widersten (1968) has recently arrived at the same opinion. He has made the interesting observation that in the larvae of cnidarians the change from planktotrophy to lecithotrophy seems generally to be accompanied by a reduction of the aboral sense organ (cf. p. 166).

different ontogeny as far as the formation of the entoderm is concerned. I wish to stress these points, since the term *planula* is often used in literature without any closer definition.

The *planula* in the special meaning of a larva with a compact entoderm developed by the immigration of scattered cells or by delamination and without a mouth is the model of the hypothetical ancestral form that has been assumed by certain researchers (Metschnikoff, Sachwatkin, Hyman, Hand, and others) and that has lately usually been called "planuloid" (for the planuloid theory see Hyman, 1940, p. 252). By subsequent changes the planuloid or its descendants should have acquired an entoderm cavity and a mouth in the same way as described for a *planula* of the latter type, and not by invagination.

I have already pointed out (1955, p. 325) that I consider the planuloid theory entirely untenable (see also below, p. 238). There is no doubt that the common ancestors of the Metazoa included a *Gastraea* form, i.e. a creature with both alimentary cavity and mouth.

This is not the place for further discussion of the question. I only want to draw attention to Fig. 57 (p. 214) which shows the occurrence of the invagination gastrula in the phyla of the Metazoa. The almost universal distribution is perhaps the most conclusive evidence for the correctness of the *Gastraea* theory.

Ctenophora

Apart from the aberrant subgroup Platyctenea, the whole life cycle of the ctenophores takes place in the pelagic zone. For the adult this mode of life must be considered as secondary. On the basis of the *Bilaterogastraea* theory (Jägersten, 1955) no other interpretation is possible. The Platyctenea have returned to life on the bottom in the adult phase of life.

Because all ctenophores possess a pelagic juvenile form there is reason to suppose that this applied also to their common ancestor. Thus here, as far as we can judge, as in the other groups of the Metazoa, a pelago-benthic life cycle is the original condition.

Again with the exception of the Platyctenea, the eggs are discharged freely into the water, where, fertilization probably takes place. The eggs are relatively rich in yolk and surrounded by an envelope in which two layers can be distinguished. The embryonic development takes place inside this envelope, and differs essentially from that of the cnidarians by being a pronounced mosaic type.

The juvenile form that is liberated at hatching agrees in all essential features with the adult within one of the subgroups, Cydippidea, and is for this reason usually called a "*cydippid*-larva". Subsequent growth in this subgroup produces only slight, gradual changes. Development is thus direct and it is therefore somewhat out of place to talk about a "larva".

Within the other subgroups, Lobata, Cestidea, Beroidea, and Platyctenea, this designation is more justified. In them the "*cydippid*-larva" undergoes considerable changes, and for this reason the term "metamorphosis" is often used in the literature.

It is commonly and probably correctly assumed that the Cydippidea are the most primitive subgroup. In all the other subgroups the adult stage is more or less altered, most of all in the benthic Platyctenea. Because the juvenile has retained its cydippid appearance we can speak of a secondary larva in these changed subgroups (see p. 229).

What then was the appearance of the primary larva common to all ctenophores? On this topic only speculations are possible. Nearest at hand is a comparison with the larvae of cnidarians, especially with those of the Anthozoa which also possess an aboral sense organ and a concentration of nervous elements underneath it. It could therefore be imagined that the pelagic primary larva of the benthic ancestors was of a *planula* type, at least in an early phase.*

In this connection it is of interest that the genus *Gastrodes* (Platyctenea) which parasitizes *Salpa* has a larva with an even covering of cilia all over the body surface. It is, however, uncertain whether or not this represents the retention of an original feature. It is possible that a reversal to an earlier condition with uniform ciliation has taken place. The ciliary organization characteristic of typical ctenophores must in any case be considered as secondary. By maintaining that the evolutionary line towards the ctenophores *could* have contained a larva of *planula* type I do not wish to assert that the group must have originated from some form of

---

* The comparison with the *planula* larva of the cnidarians is not prevented by the fact that the ctenophores as a rule move with the oral pole forward, since Remane (1956) has established that young of *Pleurobrachia* are also able to swim with the aboral pole in forward direction. The functional anterior end is thus by no means constant (cf. pp. 36 and 183).

cnidarian. *Planula*-like larvae, with or without remainders of bilaterally symmetric organization, should be more ancient than the group Cnidaria.

Neither can we form a more complete picture of the original benthic adult. If we go sufficiently far back in phylogeny, the *Bilaterogastraea* theory leads us to the assumption of a bilaterally symmetrical form that crept about by means of cilia. As far as we can judge, the ciliated plates and the tentacles characteristic of recent ctenophores were missing. These two organ systems are adaptations to pelagic life and arose or at least received their present organization in a pelagic creature.

It is difficult to tell with any degree of certainty in which way the transition to a holopelagic life has been effected. It appears most likely that the duration of the pelagic larval life was prolonged, so that eventually the larva, whatever was its organization at the time, was no longer transformed into a benthic adult. This would mean that the evolution to Ctenophora should have taken place in a neotenic way (cf. *Arachnactis albida*, p. 14; see also p. 233 *et seq.*).

The question arises when, or more correctly in which phase of the life cycle, the present pelagically adapted features (the ciliated plates, tentacles, etc.) were evolved. In my opinion there are two alternatives to choose from. The first is that they are purely larval features, which had already arisen in the ancient pelago-benthic life cycle. When holopelagic life was later adopted as a consequence of neoteny, the secondary adult became the bearer of these features.

The second possibility is that the characters were only formed in the adult once it had adopted a pelagic life. Direct development could then have arisen under the influence of adult pressure (see p. 6), involving loss of organization of the original larva.

At present it is not possible to form a definite opinion whether the pelagic organization evolved in the larval or the adult phase.

At all events the occurrence of direct development in the Ctenophora agrees with the general rule that holopelagic forms exhibit a tendency to change to this kind of development (see p. 236).

It is worthy of notice that the bisymmetry ("biradial" organization) characteristic of the ctenophores is developed as a direct result of the early blastomere cleavages. This might possibly be interpreted as an extreme effect of adult pressure, and if this view is taken it might be questioned whether the special type of early embryogenesis of the ctenophores could be considered such an important feature from taxonomical point of view as is usually done.

Tentaculata

On account of the great differences between the three groups of tentaculates I have considered it advisable to treat each group separately as far as possible.

Phoronida

All members of the Phoronida are benthic as adults, and all known life cycles, with the single exception of *Phoronis ovalis*, contain a pelagic planktotrophic larva, the well-known *actinotrocha*-larva (Fig. 2A). It should be added that nearly 30 different larvae of this type have been described, while only about 15 species are known in the adult stage. It therefore seems reasonable to conclude that a pelagic larva of the *actinotrocha* type and a benthic adult are original features of the group.

There are no essential variations in the embryogenesis of the *actinotrocha*. The developmental biology however, differs slightly in different species. For some species (*Ph. mülleri*, *Ph. pallida*) it has been established that both kinds of sexual products are freely discharged into the water, and that fertilization is external. In other species fertilization takes place in the coelom of the mother animal (*Ph. viridis*).

In several species some kind of brooding has been established. This takes place sometimes in the tentacular apparatus (e.g. *Ph. hippocrepia*), or sometimes in the dwelling tube (e.g. *Ph. albomaculata*). The larvae are, however, soon liberated and assume planktotrophic life.

It is of particular theoretical importance that in certain cases the whole development as far as metamorphosis takes place in the open water without connection with the mother animal. This has been established in at least four of the ten or so species which have been studied so far from this aspect, including *Ph. mülleri* and *Ph. pallida* (Silén, 1954). In such cases the egg is remarkably small and poor in yolk. It gives rise to a free-swimming blastula with spacious blastocoel and ciliated surface. There is no doubt that this development in the open water is more primitive than development with brooding, where a more or less complete larva is liberated (cf. p. 240). The first type must obviously have existed in the ancestor of the group.

In the Phoronida the differences between larva and adult are considerable. There can be no doubt that this is in part due to the adaptation of the two stages to their respective biotopes. Among the characters of the *actinotrocha* the large preoral lobe, with an apical nervous organ, the tentacles, and the perianal ring of cilia are particularly noticeable.

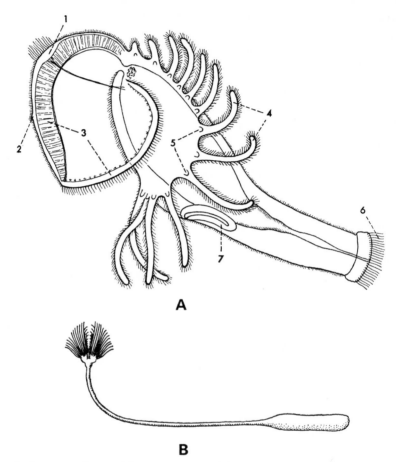

FIG. 2. A, *actinotrocha*. B, adult *Phoronis* removed from its tube. (Magnification of A over × 10 greater than that of B.) 1, apical organ; 2, sensory papilla; 3, preoral lobe; 4, first set of tentacles; 5, rudiments of the final set of tentacles; 6, perianal ring of cilia; 7, ventral invagination (metasome pouch). (Mainly after Cori, 1939.)

Because of the adaptation of the two stages to their respective biotopes the *actinotrocha* at the transition to benthic life has to undergo radical metamorphosis. Such a metamorphosis would, however, take quite a long time, and would entail the risk of an excessively long interruption in feeding, unless the development of certain adult characters had already been initiated in advance during the pelagic phase. This applies particularly to the invagination of the ventral body wall (metasome pouch) and to the final tentacles. This process of adultation, and the fact that considerable parts of the larval body (the preoral lobe, the larval tentacles, etc.) are radically discarded, provides the explanation of how the complicated metamorphosis into the adult can be accomplished in the remarkably short time of 15–25 min (Ikeda, 1901).

I use the expressions "larval" and "final" tentacles purely to distinguish between the two sets. The former are a larval character in the sense that they are only found in the larva. Before their loss at metamorphosis the final tentacles are already formed at the base (Fig. 2A), and both sets therefore exhibit the same arrangement. This agreement must not be overlooked, and Korschelt and Heider (1890, p. 1185) considered it as being due to homology.

I agree with this opinion and thus consider the two sets as in reality identical, i.e. an adult character which, because of the adult pressure, has been shifted to the larva; thus originally the larva had only one set. The replacement of this first set might possibly be explained by the assumption that after the transfer to the larva it had already assumed a function at this stage and had become so specialized that it could not be taken over by the adult at metamorphosis without some alteration. (Such specialization is obviously a purely larval character.) At first this change was small, but later on it became so extensive that it became more "convenient" to discard entirely the tentacles of the larva and to replace them by new adult tentacles. Since the latter start their development before the former are dropped, this "premature" formation has to be considered as a pre-regeneration. (It could be said that the adult character of tentacles has twice been transferred to the larva.)

This interpretation of the tentacles in the *actinotrocha* is supported also by the close agreement between the first set and the tentacles of the larva of Ecardines described on p. 47. The latter are unmistakably an adult character.*

---

* Since writing the above I have become aware that Zimmer (1964) had made the interesting observation that the "larval" tentacles of two forms of *actinotrocha*, one belonging to *Ph. vancouverensis*, the other not specifically identified, are not lost at metamorphosis, but are retained as final tentacles, there being thus only one set. This observation corroborates the interpretation given above.

The ventral invagination of the body wall (metasome pouch) which I, without hesitation, homologize with the organ for attachment (internal sac) in the larvae of Bryozoa and the peduncle of the Brachiopoda (for more detail see p. 53), is an unmistakably adult character. It differs, however, in the larva by being formed in an inverted condition. (If it were formed evaginated as in the adult, it would certainly be a hindrance to swimming.) There is of course no major difference between the two conditions, but the appearance of the body is obviously greatly altered by the extroversion and elongation of the sac during metamorphosis.

The big preoral lobe is an organ of considerable interest. It has been suggested that the epistome of the adult may be derived from this lobe. It has, however, been established that the lobe is entirely lost at metamorphosis, and that the epistome is a new formation in the transformed animal. For this reason the identity of the two organs has been doubted.

I do not think that the above reasoning justifies the rejection of the identity of the epistome and lobe. It is quite conceivable that, because of its specialized formation in the larva, the lobe could not be taken over by the adult, but had to be discarded, with subsequent regeneration as an epistome. The situation might be basically the same as in the case of the two sets of tentacles, the only difference being that the regeneration of the epistome takes place at a somewhat later stage.

There is a different reason that makes me hesitate in considering the homologization of lobe and epistome as justified; this is that there seems to be homology between the lobe and a special structure in the *cyphonautes* of the bryozoans, the mantle, which seems to be an ancient adult character. (This question will be taken up in greater detail on pp. 42 and 57.)

If this latter homologization is correct, then we can say that of the four most pronounced features of the *actinotrocha*, the preoral lobe, the tentacle apparatus, the ventral invagination, and the perianal ring of cilia, it is only the ring of cilia that can be interpreted as a purely larval character (an adaptation to the pelagic mode of life).

In the preceding account nothing has been said about one species, *Phoronis ovalis*, which exhibits a very divergent type of ontogeny (Silén, 1954) and which accordingly deserves special attention. Unfortunately important elements of the embryogenesis are still not elucidated. *Ph. ovalis* is particularly small, its length being only about 5 mm, but it nevertheless has large eggs, rich in yolk; their diameter is about 125 μm. For the sake of comparison we can mention that the eggs of *Ph. mülleri*—the body length here reaches almost 50 mm—measure only about 60 μm.

The wealth of yolk and divergent brooding in *Ph. ovalis* are related to a considerably altered development. The eggs are laid in the tube, and there the embryos remain until gastrulation has taken place. The process of gastrulation is not known in any detail, but presumably represents an altered invagination. The organization of the body is already distinctly bilaterally symmetrical at the *gastrula* stage, and it seems that the invagination should take place on the future ventral side and not at the posterior pole of the blastula. Unfortunately the polarity has not been definitely settled.

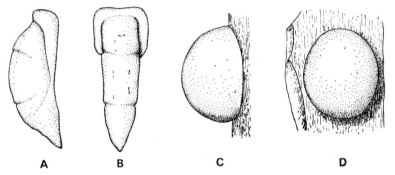

FIG. 3. *Phoronis ovalis*. A and B, larva creeping on the substratum, lateral and dorsal view; C and D, larva just settled and with all external differentiations obliterated. (After Silén, 1954.)

Here we touch on a question which is theoretically very important. If it is true that invagination does take place on the future ventral side, this might be interpreted as support for the *Bilaterogastraea* theory or as an extreme effect of adult pressure. Indications of similar conditions also seem to occur in other species. According to Selys-Longchamps (1907) the blastula is already bilaterally symmetrical in *Ph. mülleri*, and the invagination takes place on its ventral side; also Schultz (1897) and Masterman (1900) supply information and illustrations which point in the same direction. A detailed investigation which should also be extended to other groups is highly desirable.

The larva of *Ph. ovalis* differs radically from the *actinotrocha* (Fig. 3), particularly in the ciliary covering which is uniformly distributed all over

the body, and by the total lack of tentacles, alimentary cavity and functional mouth and anus. It is completely lecithotrophic and does not
assume a true pelagic life. It moves about freely only for some few days,
first swimming near the bottom, then gliding over the substratum somewhat like a turbellarian. Finally it settles down and is transformed into the
adult. No details of the transformation are known, but Silén speaks of ''an
almost direct development''. No shedding of larval parts should occur.
We obviously have here an altered and simplified ontogeny, implying
among other things total loss of the tentacles. (The rich supply of yolk
in the egg is a necessary condition for this change.) In all probability *Ph.
ovalis* originally had an *actinotrocha*. Silén is even of the opinion that the
larva exhibits certain retained traits of the *actinotrocha* organization. The
meaning of this expression is, however, not quite clear. To my mind the
larva of *Ph. ovalis* appears so strongly altered that it is difficult to trace its
original structures.

Theoretically it can of course be imagined that in the particular line of
evolution leading to *Ph. ovalis* the formation of the tentacles has never been
shifted to the larva (cf. discussion on p. 56), but to me this possibility
appears less probable. The absence of tentacles in the larva more likely
depends on the wealth of yolk in the egg. What has happened is probably
a ''retardation'', a delay in the formation of the tentacles.

As a general result of our review of the situation in the Phoronida it
can be asserted without doubt that a pelagic larva and a benthic adult can
be considered as original for the group. In spite of the divergent larva in
Ph. ovalis the assumption must also be justified that the ancestral form
common to all species had a planktotrophic *actinotrocha*. As far as can be
judged the following also seem to be original for the group: small egg,
poor in yolk and discharged freely into the water; external fertilization;
ciliated coeloblastula; and gastrulation by invagination.

We may finally recall the interesting fact that the longitudinal axis of
the *actinotrocha* is at right angles to that of the adult. This is because the
adult, on account of its sessile mode of life, has undergone a radical
transformation of the body. The axial organization of the *actinotrocha*
must therefore be designated as an ancient adult character.

Bryozoa

The life cycle consists of the two common phases, a pelagic larval
phase and a benthic—in this case sessile—adult phase. There are no ex-

ceptions to this pelago-benthic life cycle, even though the free-swimming period of the larva is usually of very short duration.

In the Bryozoa we can distinguish two main categories of larvae which are markedly unlike in their conditions of development and mode of life; these are the planktotrophic *cyphonautes*-larvae (Fig. 4A), which pass their entire development in the pelagic zone, and lecithotrophic forms (Fig. 4B and C) which are all brooded, usually in different types of special brood chambers. Morphologically, however, these two categories are connected by intermediate forms (see below). It might be mentioned that the designation lecithotrophic is not quite correct in all cases, since certain larvae are developed from eggs with little yolk and during embryogenesis are supplied with nutriment from the mother animal (see below).

There are lecithotrophic larvae in the vast majority. In the Phylactolaemata and the Cyclostomata the *cyphonautes* is entirely unknown and must be non-existent. In the Ctenostomata it has been observed in species belonging to well separated genera, viz. *Farrella*, *Hypophorella*, and *Alcyonidium*. In the last case there exists the remarkable situation that *A. albidum* possesses a typical *cyphonautes*, while at least in *A. polyoum* the larva is of the lecithotrophic type. In the Cheilostomata the *cyphonautes* is known only in the mutually closely related genera *Membranipora* and *Electra*. To this it should be added that both in the Ctenostomata (*Flustrella hispida*) and in the Cheilostomata (*Scruparia chelata*), larvae are found which combine a slightly altered *cyphonautes* appearance with lecithotrophy and development within the mother animal. In the former case even the typical lateral shells occur. It is almost a matter of taste whether or not such a larva as that of *Flustrella* should be called *cyphonautes*.

Here we touch on a most important point. While the planktotrophic larvae represent a very uniform type, the larvae which are brooded exhibit in several respects considerable variations. They constitute a whole series of forms from fairly *cyphonautes*-like types to the strongly aberrant larvae of the Phylactolaemata (see below).

In species with planktotrophic larvae the eggs are small, poor in yolk, and freely discharged into the water. No exceptions to these characteristics are known, and nothing indicates the occurrence of brooding in the past. It is obvious that the lengthy pelagic life of the larva and its planktotrophy are connected with the nature of the egg. Here planktotrophy is quite simply a necessary condition for the development into a mature larva.

In forms with lecithotrophic larvae the eggs are relatively large and rich in yolk. On leaving the mother animal the larvae are fully developed and for this reason pass only a brief pelagic period, often perhaps just sufficiently long to find a spot suitable for the transformation into the adult. There is undoubtedly a correlation here between the properties of the egg and the mode of life of the larva. A fairly large amount of yolk is in fact a necessary condition for lecithotrophy.

It has already been pointed out that the eggs of certain bryozoans with brooding, especially the Phylactolaemata, are small and poor in yolk. This is related to the fact that the embryo during development is supplied with nutriment from the mother animal, either by placental structures or otherwise. It is obvious that these arrangements, which are certainly secondary, are necessary conditions for the development of these small eggs into larvae. During evolution, the amount of yolk in the egg has been reduced concurrently with the increasing refinement of methods of obtaining nutriment from the mother animal. This means of nutrition, as well as brooding in general, has probably arisen independently several times within the Bryozoa, e.g. in *Sundanella* (Braem, 1939) and *Bugula* (Marcus, 1938) of the Gymnolaemata and in Phylactolaemata.

It is thus possible to distinguish among the bryozoans three different situations concerning the properties of the eggs and the conditions connected with them:

1. Small mass of yolk, freely discharged eggs, planktotrophy, and lengthy pelagic life;
2. Abundance of yolk, brooding, lecithotrophy, and short pelagic life;
3. Small mass of yolk, intimate brooding with supply of nutriment during embryogenesis, and short pelagic life.

There can be no doubt that the changes have taken place in this order. This is corroborated, *inter alia*, by the properties of the digestive tract.

FIG. 4. Diagrams showing four of the most important main types of larvae within the Bryozoa, all orientated in the same way. A–D, pelagic stage; A^1–D^1, shortly after attachment to the substratum. A, *cyphonautes*; B, lecithotrophic form with remnants of the digestive canal (*Cheilo-Ctenostomata*); C, without remnant of the digestive canal (*Cyclostomata*); D, larva of *Phylactolaemata*. 1, pyriform organ; 2, apical organ; 3, stomodaeum; 4, mesenteron (or rest of it); 5, adhesive organ; 6, shell (in A^1 about to be lost); 7, mantle lobe; 8, anlage of polypid.

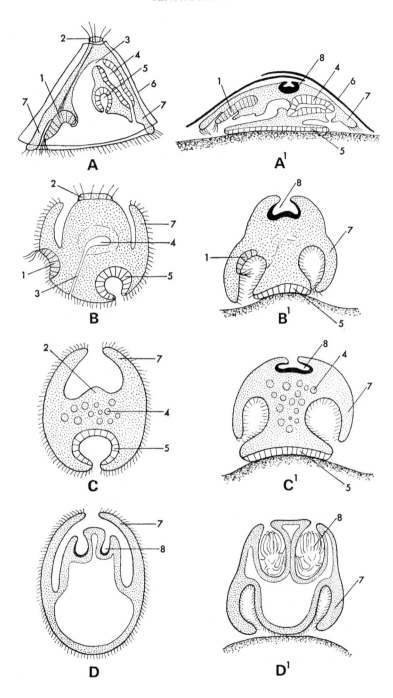

In *cyphonautes* it is of course fully developed. In the lecithotrophic larvae on the other hand, there is the common feature that a continuous digestive canal is missing. In some only the posterior part and the anus are missing (e.g. *Bulbella*, according to Braem, 1951). In others only the mouth and the oesophagus is retained (*Tanganella*, according to Braem, 1951), while in the majority of cases the remnants of the digestive tract are entirely without lumen. In such cases the entoderm can often still be traced as a compact mass of cells with yolk granules. Finally, in the case of the Phylactolaemata it is not certain that an entoderm is ever formed (see below).

The foregoing evidence very strongly suggests that *cyphonautes*, although it is in the minority, must be considered the most ancient type of larva, and that those larvae which are brooded are derived from it. This is also the most widely held view for which more supporting evidence will be given below.

A conflicting idea has, however, been proposed by Silén (1944a). The study of the different organs for brooding within the group has led him to the opinion that these structures are homologous and, consequently, that the brooded lecithotrophic larvae are more ancient than the planktotrophic and more uniform *cyphonautes*-larvae. The latter should have originated independently from the former in several places in the evolutionary lines within the group.

I shall not examine Silén's idea more closely, because I consider it impossible. It has also been attacked by Brien (1960, p. 1273). Within the entire animal kingdom we do not know a single example of a larva which, after having lost the habit of planktotrophy and having had its alimentary system reduced or eliminated, has reverted later on to the original conditions. We know, on the contrary, numerous certain parallel cases of secondarily evolved brooding and the accompanying alteration of the larvae, especially of their alimentary tract. (Or the larvae may have become replaced by direct development.)

Here I want to recall the lecithotrophic larva of *Phoronis ovalis* (see p. 29 and Fig. 3) which is considered by Silén himself as an altered *actinotrocha*. This alteration obviously constitutes a parallel case to that of the lecithotrophic larvae of the bryozoans.

As has already been mentioned both *cyphonautes*-larvae and the altered forms are found occurring together within various families and genera. This makes it obvious that the change has been effected independently in several evolutionary lines. It is worthy of notice that the result has nevertheless been almost the same in several cases. Most likely the explanation can be found in the fact that the original form, the *cyphonautes*, is such a uniform larval type and that the new environment (the body of the mother animal) has always been the same.

In all cases the incipient alteration must have consisted in a gradually increasing mass of yolk in the egg, which in turn influenced the alimentary tract. The transformation of the latter into being simply a container of the yolk could continue in these lecithotrophic larvae without entailing any complications in the continuation of embryogenesis, since in the recent bryozoans the alimentary system of the adult is never derived from that of the larva, but is formed from the ectoderm in the course of metamorphosis.

Although, in general, the situation can be considered as elucidated, new detailed investigations appear desirable. Such investigations might not only supply new views upon parallelisms in evolution, but might also lead to valuable results concerning the taxonomy of the bryozoans. New studies ought also to be made of the different organs for brooding within the group although it can now be safely asserted that these have arisen independently in several places.

We must return once more to the profoundly altered larva of the Phylactolaemata. Its organization is roughly represented in Fig. 4D. It is a somewhat elongated cyst with two layers and ciliary covering all over the free surface. At one of its ends one or several polypid buds are developed during embryonic life. This early formation of polypids which I consider as an effect of adult pressure (see below, p. 45) has left such a mark on this larva that the development has occasionally been designated as a direct one and the larva as a swimming "primary colony". This view is held, e.g., by Kaestner in his "Lehrbuch der speziellen Zoologie" (1963). The end with the polypids is enveloped by a mantle which issues from roughly the middle of the body and which only leaves a very small opening at the extreme end. At the opposite pole the ectoderm contains a concentration of nervous elements and sensory cells.

No agreement exists as to which is the front end and which is the hind end in this larva or, more correctly, what corresponds to the apical (animal) and oral (vegetal) pole, respectively, of the other bryozoan larvae. Braem (1897), Marcus (1938, 1939), and others are of the opinion that the pole at which the polypids bud off is the vegetal, while Barrois

(1882) believes it to be the animal pole. Brien (1953) doubts the opinion of Braem and Marcus, but does not enter into any criticism.

The principal argument for considering the polypid pole as the vegetal is supplied by Braem (1897), who believed that in *Plumatella* he could observe an invagination in this pole. (The entoderm formed should, however, become dissolved almost immediately.) This interpretation can be supported also by the concentration of nervous elements and sensory cells at the opposite pole, and by the fact that during the swimming of the larva this ''apolypidal'' pole is usually directed forward. The latter argument, however, loses some strength, since firstly the larva is known to change its direction of movement frequently and to swim with the polypid pole foremost (Brien, 1960), and secondly because other larvae have been found able to move backwards (see, e.g., p. 183).

Decisive arguments against Braem's and Marcus' concepts are supplied by the organization of the mantle and its behaviour during metamorphosis, and by the fact that the larva attaches itself with the pole opposite to that bearing polypids. Here there is complete agreement with the larvae of the Gymnolaemata. A comparison of the Phylactolaemata-larva with almost any lecithotrophic larva having a moderately well-developed mantle within the former group shows that in either case the mantle envelops that end of the body at which the polypids are developed, and that the surface of the mantle which in this position is the inner one is devoid of cilia, while the outer one is ciliated. Furthermore, in both cases the mantle is similarly inverted towards the substratum at metamorphosis (Braem, 1897). There is thus no valid reason for the assumption that the mantle of the bryozoan larva is not always and everywhere the same organ.*

The differences which exist between the common type of lecithotrophic larvae with a mantle in the Gymnolaemata and those of the Phylactolaemata are involved mainly with the fact that in the latter so many organs have been eliminated (the apical organ, the pyriform organ,

---

* Remarkably enough Brien (1953) has arrived at a different result concerning the metamorphosis of the Phylactolaemata. According to him the larva is already transformed during its free-swimming life, and the mantle is not inverted in the way described by Braem and others. One wonders whether Brien's results might not be attributable to the fact that his larvae found no natural substratum for attachment. We know now that for metamorphosis many larvae require the substratum from the biotope of the adult. In any case, Brien's results in no way interfere with the homologization of the poles presented here.

the adhesive organ, and the alimentary system). An additional difference is that the development of the polypids starts already during embryonic life. Also this feature is of course a secondary phenomenon.*

I have thus arrived at the conclusion that in the larvae of the Phylacto-laemata the pole at which the polypids are formed corresponds with the apical pole of the other bryozoan larvae, and that the opposite pole represents the transformed oral pole. The opposite interpretation adopted by Braem and Marcus is morphologically impossible. I must stress this point even if Braem should be right in his assertion that in the Phylacto-laemata an invagination actually takes place at the pole of the blastula that is directed towards the point of attachment of the embryonic pouch. If this is correct, we should have to assume either that the "pseudoblastula" resulting from the disappearance of the entoderm undergoes a re-polarization (perhaps by uneven growth) or else a reversal. The sensory cells and the concentration of nerves, established by Marcus at the oral pole, might very well be new formations explained by the fact that the larva swims preferentially with this pole forward. (I wish to remind the reader of the existence of pygidial eyes in the majority of the members of the family Sabellidae among the Polychaeta, which have been formed in connection with the adoption of backward locomotion.)

The elucidation of the polarity in the embryos and larvae of the bryo-zoans is of additional importance because Marcus (1939) makes his erroneous interpretation of the poles in the larvae of Phylactolaemata the basis for his fusion of Entoprocta and Bryozoa. I may return to the question in a future paper (see, however, also p. 115).

Lemche (1963) also starts from an erroneous interpretation of the poles, when he makes his strange attempt to split the group Bryozoa and to derive the Phylactolaemata "directly from some coelenterate stock of rhizo-stomean affinities" (!).

Having arrived at the result that the larva of the Phylactolaemata can be joined to the series of types of more or less modified larvae which occur among the other Bryozoa, and that all of them can be derived from *cyphonautes*, I do not necessarily wish to imply the conclusion that the

---

* Perhaps the adhesive organ has not been altogether lost? Braem (1897) has reported the occurrence of gland cells in the apolypid pole of the *Plumatella* larva. Here it might be a question about remains of the gland cells belonging to the adhesive organ which is in general strongly developed in the Gymnolaemata. I should not exclude the possibility of interpreting the conditions in the Phylactolaemata by assuming that their organ for attach-ment has been permanently turned inside out.

primeval bryozoan, i.e. the last common ancestor of all bryozoans, had a larva that exactly resembled the recent *cyphonautes*. It appears nevertheless very probable that the larva of the primeval bryozoan in essential respects was quite close to *cyphonautes*. I shall return later to the problem where in the phylogenetic tree of the tentaculates certain *cyphonautes* characters have made their appearance (see p. 60).

The following comment should be added on the subject of the mantle of the recent bryozoan larvae. It is remarkable that this part of the body is also retained in the altered larvae, though often in strongly reduced condition, and that its behaviour during metamorphosis is essentially the same everywhere. Unlike the *cyphonautes*, however, it is as a rule markedly altered in those larvae without shells. Unless it is altogether too feebly developed it is folded up towards the apical pole. In certain cases, and most completely in the Phylactolaemata it includes the apical half of the body (Fig. 4B and D). During metamorphosis the conditions existing in *cyphonautes* are in principle re-established. This is effected by a folding back of the mantle against the substratum. In certain cases at least the mantle then contributes to the attachment of the body.*

For *Tanganella* (related to *Paludicella*) Braem (1951) reports an observation that is of interest here. The larva is of the common lecithotrophic type with upturned mantle (Fig. 5B and C). In a fairly late stage of embryogenesis the mantle is, however, directed downwards, and for a short period the margin is even entirely closed ventrally (Fig. 5A). Apart from this closure the agreement with *cyphonautes* is striking (cf. Fig. 4A). To me it seems very likely that this represents a reminiscence of a time when the larva was still a typical *cyphonautes*. Remarkably enough, the resemblance with *cyphonautes* returns at metamorphosis (Fig. 5D).

As in the case of other primary larvae the following kinds of features can be expected in *cyphonautes*: (a) purely larval characters (adaptations

---

* The use of the mantle, in some cases, as an organ of attachment should be more closely examined. If the statements on this point are correct, a certain analogy exists with the way in which the larvae of certain entoprocts are said to attach themselves, that is by means of the atrial margin carrying the prototroch. There is, however, no question about anything deeper than an analogy. The resemblance between the two groups in this respect, as in so many others, has no phylogenetic basis. The original adhesive organ of the bryozoans is an entirely different structure (see p. 47), while the ancestral forms of the entoprocts on their passage to a sedentary life attached themselves with the preoral portion of the body (see p. 109).

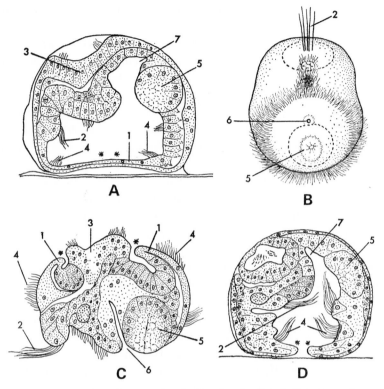

Fig. 5. Ontogenetic stages of *Tanganella* (closely related to *Paludicella*). A, median section through the embryo inside the mother animal. Note the general agreement with a *cyphonautes*. Between asterisks, temporary fusion of the margins of the mantle. B, fully developed larva in ventral view. C, same in median section. The asterisks mark the same points as those in A. D, larva attached to the substratum and undergoing metamorphosis. The asterisks indicate the position of later coalescence. 1, mantle lobe; 2, tuft of sensory hairs on pyriform organ; 3, apical organ; 4, corona; 5, adhesive organ; 6, mouth; 7, remnant of gut. (After Braem, 1951.)

to pelagic life), (b) recent adult characters, and (c) ancient adult characters.

Little needs to be said about the purely larval characters. Of them, the ring of cilia at the margin of the mantle is the most noticeable. It has frequently been asserted that this ring, the corona, should be homologous with the prototroch of the *trochophora*-larva. This interpretation is uncertain, but perhaps not impossible (for this point see p. 64).

Of the other characters occurring in *cyphonautes* the following are of special interest: the shells, the mantle, the pyriform organ, and the adhesive organ.

Let us first examine the shells. What is their nature? They can under no circumstances be considered recent adult characters, since no trace of such structures is found in the bryozoan colonies. Neither do they obviously represent larval characters. Shells occur only in some few genera within two separate subgroups. If they were of special value in the pelagic environment, e.g. for protection, there would be reason to expect them to occur more commonly. Everything suggests that the shells like many other features of the *cyphonautes* have been eliminated in the great majority of the larvae, which are all more or less altered. It must be kept in mind that shell structures which can be interpreted as purely larval (pelagic) characters are not found in the primary larvae of other groups either.

There is almost no doubt that in *cyphonautes* the shells constitute something of a disadvantage, both by their weight and because they restrict the extension of the ciliated area (in the majority of the altered bryozoan larvae this area is considerably greater).

A study of the swimming of the *cyphonautes* often creates the impression that it is very laborious. As soon as the movements of the cilia cease the larva sinks to the bottom of the observation dish just like a *veliger*-larva provided with a shell. Once on the bottom it remains motionless for some time as if recovering strength after its exertion.

The foregoing evidences suggest strongly that the shells of *cyphonautes* do not represent an exclusively larval character, but are, like the shell in the *veliger*-larvae (see p. 119), an adult character that has been shifted to the pelagic phase of life. Now, as has already been said, the adult bryozoans have no such lateral shells, even if the surface of their body is usually strongly cuticularized in other ways. Thus everything suggests that the shells of *cyphonautes* are an ancient adult character, an immensely ancient heritage from one of the many ancestors in the phylogenetic line leading to the bryozoans. In other words, *cyphonautes* is in this respect, as in many others, more conservative than the adult which has become strongly altered in conjunction with the transition to a sessile colonial life.

In case anybody should wish to assert that the *cyphonautes* shells, being a kind of cuticular formations, cannot be considered of any phylogenetic significance, or that they should be interpreted as a recent adult character, on the basis that many Gymnolaemata are strongly cuticularized as adults,

I want to stress that such reasoning appears to me absolutely unfounded. Morphologically distinct cuticular shells—the *cyphonautes* shells are fully equivalent to e.g. the shells of ostracodes or mussels, having like them an adductor muscle—are as useful for comparative studies as "soft" characters and organs. It has also to be noted that the *cyphonautes* shells are situated on a preoral (dorsal) part of the body, while the strong cuticula of the adult covers a postoral (ventral) portion. The shells, furthermore, tend to be reduced as far as can be judged, being very thin in certain *cyphonautes*-larvae and absent in others. Reduction is not the general rule for recent adult characters of larvae, but is quite frequent among ancient ones (see p. 222).

By describing the *cyphonautes* shells as an (ancient) adult character I do not want to deny categorically the possibility that they might have some function. (If such a function arose after the shifting of the shells to the pelagic phase, it is obviously a purely larval feature; cf. p. 223.) But, as already indicated, there is nothing that points to any function of the shells in *cyphonautes*. It might of course be imagined that they have a protective value, but this is only a guess. Other larvae do quite well without any shells.

According to the above reasoning, the situation in bryozoans with retained *cyphonautes* is the same as in many transformed gastropods, where the adults have lost every trace of the shell, while the larvae still possess a marked one. Just as the *veliger*-larva of a gastropod, which is so strongly transformed that it would be impossible to place if in the proper group, unless its larva were taken into account (e.g. *Entoconcha*), tells us that the ancestor had a shell also in the adult stage, in the same way the *cyphonautes* supplies evidence of an early bryozoan ancestor (cf. also p. 122).

We can, however, follow this train of thought still further. The ancestor at which we have now arrived cannot very well have been the "primeval bryozoan", i.e. the *last* common ancestral form of all bryozoans, since this must have been colonial and therefore had in all probability already lost the shells of the adult stage. (There is, in fact, no more reason to assume that the primeval bryozoan possessed lateral shells than that such shells were found in some as yet unknown recent colonial bryozoan.) The ancestor in question must be still older, and it must have been solitary. It was the shells of this solitary ancestor which, because of adult pressure were already formed in the pelagic primary larva. This change ("acceleration") must have taken place while the adult still retained its lateral shells. Thus it is reasonable to assert that the *cyphonautes*—

or at all events a larva with shells—is older than the bryozoans. I shall return to this question later (see p. 59 *et seq.*).

Having arrived at the conclusion that the *cyphonautes* shells must be considered an ancient adult character we are compelled to consider also their matrix, that is the underlying mantle, as the same kind of feature. As in the Mollusca mantle and shell together must be conceived as a unit. (As has already been pointed out, the ciliation of the mantle margin, the corona, is, however, a separate feature—a larval one.)

The so-called pyriform organ, which is characteristic for the majority of the bryozoan larvae, has been studied by a number of scientists, most throughly by Kupelwieser (1905). He has found that in the mature *cyphonautes* the organ can be extended like a tongue while the larva moves over the substratum, and he supposes that the organ then serves as a tactile organ. This seems to be indicated by the probing movements of the long cilia ("plumet") at its apex.

Using large larvae from southern India, most probably belonging to *Membranipora*, I have been able to verify that the organ is extended in different directions while the animal is creeping about, and I have furthermore found that locomotion over the substratum is effected just by the organ. In reality the creeping takes place, at least for the major part, by means of the cilia upon the lower surface of the tongue (Fig. 6). As far

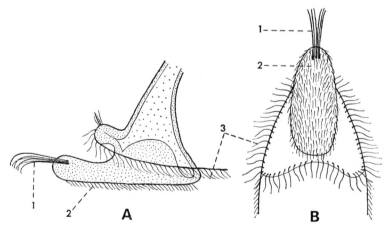

FIG. 6. Diagrams of the pyriform organ of a *cyphonautes* while used as an organ for creeping. A, seen from the left (partly after Kupelwieser, 1905); B, ventral view; 1, tuft of long cilia (with sensory function ?); 2, ciliation of the ventral side of the organ; 3, corona of the margin of the mantle.

as I have been able to establish, Kupelwieser is wrong in his assumption that the creeping locomotion is effected exclusively by means of the corona.

By establishing that the pyriform organ is a locomotory organ I of course do not exclude the possibility that it might also have a sensory function. On the contrary, I consider this probable because of the movements of the "plumet" of long cilia, but a function of this nature is of course much more difficult to prove than that of locomotion.

In the majority of the lecithotrophic larvae, too, the pyriform organ is retained, though in a more or less reduced condition, but there the organ seems to have lost its locomotory function. The tuft of long cilia is, however, still well developed in many forms, and its movements are still the same. This appears, e.g. from Silén's (1944b) account of the larva of a species of *Callopora*. It is remarkable that Silén gives no reason for saying that the kind of movement of the long cilia "forbids . . . that they would serve as tactile organs". Not only Kupelwieser but also Braem (1951, p. 29) holds the opposite view. Silén considers it probable that the pyriform organ serves for the ingestion of food. He is led to this opinion by the type of movement of the cilia. This is not, however, a reliable argument. There is no reason to discard the opinion that this larval type (see Figs. 5B and C) in its short period of free living is entirely lecithotrophic.

I now come to the important question of how the creeping organ of *cyphonautes* should be interpreted from the phylogenetic point of view. We have the choice of considering it either a purely larval (pelagic) structure or as an ancient adult character. (It is obviously not a recent adult character.)

I want to say immediately that I find myself compelled to consider the creeping organ as an ancient adult character. Only its function in relation to the substratum of the bottom has been established with certainty. In this respect its features are entirely analogous with those in the larvae of Entoprocta and Sipunculida (see pp. 105 and 150). Particularly striking is the similarity of function to the retractile foot of the larvae of the Entoprocta, where long probing cilia also are found in the anterior end (Fig. 24, p. 103). Any thought of homology is, however, made quite impossible by the fact that in the Entoprocta the creeping organ occupies a postoral (ventral) position.

For the sake of completeness it should be mentioned that Kupelwieser imagines that the pyriform organ of *cyphonautes* may also have a function to fulfil while the larva is swimming. He thinks that it might transport particles of food into the vestibulum (atrium) for further conveyance to the mouth. If this really is the case, it might be interpreted as an argument against my interpretation. Atkins (1955) has, however, found that feeding takes place in an entirely different way—without the collaboration of the pyriform organ.

From the above we can see that the discussed situations in *cyphonautes* point to the conclusion that in the evolutionary line leading to the bryozoans there has existed an ancestral form which in the adult stage crept upon the substratum by means of a ciliated preoral part of the body (see further p. 61).

It is worth stressing that this part of the body actually has a function in the larva at the time when it moves to the location where metamorphosis is to take place. In this respect there is a similarity with the creeping organ of at least some annelid larvae (p. 172; see also p. 229).

There is an important adult character in the bryozoan larvae, which, although it is very ancient, should nevertheless be interpreted as a recent feature; this is the adhesive organ (internal sac). This organ must be derived from the common ancestor of all tentaculates and should thus be homologous both with the ventral invagination (metasome pouch) of the Phoronida and the stalk (peduncle) of the Brachiopoda. In Bryozoa and Phoronida the structure is developed as an invagination, between mouth and anus, on the ventral side of the larva. In the Brachiopoda its location is basically the same, although further back. For all three groups it is evident that the organ is formed in the larva in preparation for the adult phase of life.

The adhesive organ of the bryozoans functions when the larva settles. In other respects there are said to be considerable differences in the different forms. In *Tanganella* it has no further function, but is obliterated. In others it may form all or part of the final epidermis (see Braem, 1951, p. 29). (See p. 52 *et seq.*)

It remains for something to be said about the metamorphosis of the larva into the adult and the formation of colonies. Because of the strong difference between larva and adult the metamorphosis is very far-reaching. Originally, as in most other groups, consisting of a transformation of the larval body while retaining at least its more essential parts, e.g. the alimentary system, it finally involved the disintegration of all inner organs of the body and the development of the polypid by budding-off from the ectoderm. This recent metamorphosis can under no circumstances be considered as original.

So far as altered metamorphosis is concerned the bryozoans have gone one step further than the entoprocts; the original situation, i.e. transformation of the entire larval body into the adult, is nowhere retained. For the entoprocts, I consider it evident that the newly-discovered formation of buds in the interior of the larva (see below, p. 106) is the same phenomenon as the vegetative propagation in the adult that has been known for a long time. This propagation, a common result of a sedentary mode of life, was first evolved in the adult, but was extended later to the larva. According to my opinion the principle is the same in the bryozoans. Although the two groups have no other phylogenetic connection than that they are both protostomians, adult pressure has led to similar results in their widely separated evolutionary lines.

This reasoning has led me to a definite attitude to an old difference of opinion among students of bryozoans, as to whether or not the formation of the polypid at metamorphosis is the same process as occurs later in the colony, i.e. a simple budding process. In spite of the similarity, Korschelt (Korschelt and Heider, 1936, p. 439) considers "die Polypidanlage der Larve als eine Vervollständigung des Individuums zur Ausbildung der ihm fehlenden Teile beim Abschluss der Metamorphose". I do not find this explanation satisfactory. Admittedly it appears in the course of the ontogeny as if the formation of buds in the metamorphosing larval body were a regeneration of lost organs and parts of the body, but from the phylogenetic point of view I am unable to visualize things in this way. It is impossible to believe that, phylogenetically speaking, the larva had started to "degenerate" before the new type of development was fully established. In other words, budding (regeneration according to Korschelt) in the larva appeared in the line of evolution prior to "degeneration". It seems to me inconsistent to consider the budding in the larva as regeneration and an exactly similar budding in the adult as propagation, a different thing. In my opinion we are concerned with identical processes, viz. vegetative propagation (formation of colonies) by budding. Such propagation is, however, characteristic for the adults in sessile groups, but not for larvae. In all probability the situation is as follows: in the bryozoans (as well as in the Entoprocta) the larva continued to transform itself into the adult in the "ordinary" way, i.e. by retaining the alimentary system, even at a time when the adult had already become colonial by vegetative propagation (cf. *Pedicellina*, p. 109). The original course of ontogeny was eliminated from the phylogeny as no longer necessary once the ability to form buds had been extended ("accelerated") to the metamorphosing larva. Thus application of the principle of adult pressure supplies an elucidation of the peculiar metamorphosis of the bryozoans. Except in the Phylactolaemata the formation of the polypid buds has, however, not been transferred to the free larval phase of the life cycle.

The evaluation of colonial organization and the changes connected with it involved a strong increase of the divergence which already existed between larva and adult. This increased divergence has been retained up to the present time. It has not been reduced in the many cases where the *cyphonautes* has been changed into a lecithotrophic larva. This change is obviously not due to adult pressure. The present adult features of the bryozoans are in general unfit for development at the larval stage.

Brachiopoda

All members of this group are benthic as adults and have juvenile forms which are pelagic at least for a short time.

In the Ecardines both kinds of sexual products are freely discharged into the water, where fertilization and development take place. In some Testicardines, on the other hand, altered conditions have been established. The fertilized eggs are somehow retained in the mantle cavity or even in the nephridia (*Argyrotheca*), and the larvae reach different stages of development inside the mother animal before swarming out.

In both subgroups the cleavage of the eggs is holoblastic and practically equal, and gives rise to a coeloblastula which by invagination is transformed into a gastrula.

As in the Phoronida it appears at least in some cases as if the invagination does not take place in the posterior pole but in the side destined as ventral (see e.g. Plenk, 1915). If this is the case, the situation may be of theoretical importance (see above, p. 29).

Considerable differences exist between the two subgroups as to the pelagic phase of the life cycle. In the Testicardines the larva is lecithotrophic in all known cases, and the pelagic phase is fairly short. In the Ecardines, where unfortunately a closer investigation of the embryogenesis has only been carried out on *Lingula* (Yatsu, 1902), planktotrophy has been established both in this genus and in some larvae of related forms.

From a general point of view the development of *Lingula* is of considerable interest. In the course of embryogenesis inside the thin egg membrane the body has already roughly assumed the shape of the adult. Thus, when the membrane is shed, mantle lobes, shells, and lophophore can be clearly distinguished, and after a short time the peduncle is formed, but this is, however, coiled up between the valves of the shell during the pelagic phase. The mantle lobes are directed forward as in the adult. The tentacles are strongly ciliated and serve both for swimming and feeding. The food consists of diatoms and other micro-particles, exactly as in the benthic adult. Thus the transition to the sessile phase does not entail any radical changes.

In the course of the evolution of *Lingula* the initial development of several recent adult characters (tentacles, mantle lobes, shells, setae, peduncle, etc.) have been shifted to the pelagic larva. This has involved such great changes that morphologically the development must almost be considered as direct. Nothing seems to remain of the purely larval characters. In spite of this one speaks of a "larva", and this can be justified by its pelagic life. It is only the retention of the pelagic life that distinguishes the development in *Lingula* from typical direct development. Since its numerous adult characters fit its pelagic life or at least do not form any direct obstacle, it seems that the pelagic phase might even have been quite prolonged.

It seems as if the development of *Lingula* might on the whole be representative of the Ecardines, since one occasionally encounters other larvae of the group which in external appearance closely resemble *Lingula*. In this context I want to mention some observations made on a larval form of which I have found some ten specimens in the Florida current in February, 1963 (Fig. 7).

What is most striking in this larva is the fact that the tentacles (cirri) are not situated on a distinctly developed lophophore as in the adult, but are arranged in a way that almost exactly coincides with the situation in an *actinotrocha* (cf. Fig. 2). When the anterior end is withdrawn between the mantle lobes the general likeness with an *actinotrocha* becomes quite striking, provided no notice is taken of the mantle lobes and the shells. Just as in the *Phoronis*-larva, the tentacles are arranged in decreasing size in an arc. Ventrally this arc forms a transverse row which then runs obliquely upwards and forwards on either side with the result that its ends almost meet dorsally, roughly above the mouth. At their bases the tentacles are mutually joined with the result that they might be described as arising from a low ridge, backwardly directed on the ventral side (Fig. 7B).

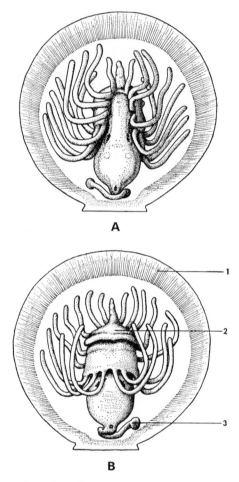

FIG. 7. Pelagic *Ecardines* larva from the Florida current, seen (A) from the dorsal and (B) from the ventral side. 1, setae at margin of mantle; 2, preoral lobe (anterior lobe); 3, peduncle. (From preserved and contracted material. Diameter of the shell *ca*. 1 mm.)

The preoral portion of the body (preoral lobe) agrees on the whole with the same structure of the *actinotrocha*, but is relatively smaller. This preoral lobe bears an anteriorly directed papilla, rather like a tentacle but with a stouter base. The papilla is rather more ventrally situated than the foremost (uppermost) pair of tentacles (Fig. 7).

These larvae were often found to be swimming about in the dishes. During swimming the two valves of the shell are opened sufficiently to allow the protrusion of the tentacular apparatus in anterior direction. This is made possible by the fact that the part of the body behind the tentacles is highly extensible. Swimming is effected by the activity of the cilia on the tentacles.

Before making an attempt to explain what has been established I shall give a brief account of the Testicardines. Here the morphological features are in part essentially different in spite of the fact that some recent adult characters are already developed during the pelagic phase of life. Of these the most noticeable are the mantle lobes bearing setae. The lobes, which are devoid of shells, are formed in backward orientation.

It must be stressed here that the mantle and shell of the brachiopods have nothing to do with the structures bearing the same names in the *cyphonautes* of the bryozoans. I cannot see any basis here for homologization. (For discussion of the phylogeny of the mantle structures see p. 57.)

The larvae of the Testicardines have neither tentacles nor mouth opening. These are not formed until metamorphosis, after the transition to sedentary life. All known larvae of this group are lecithotrophic.

The anterior portion of the body is covered with cilia and as a rule sharply separated from the middle portion by a circular constriction. At least in some forms, e.g. *Argyrotheca*, this "anterior lobe" is umbrella-shaped and carries strong cilia at its margin. The central part of the anterior lobe can be differentiated into an almost head-like elevation provided with eyes (*Lacazella*).

It might appear probable that such an umbrella-shaped anterior lobe like that in the larva of *Argyrotheca* as represented by Kowalevsky's illustration, reproduced in Fig. 8, is a purely larval character, which has arisen as an adaptation to the pelagic mode of life. There can be no doubt that the lobe is an efficient organ for floating, but this remark does not answer the question about its phylogenetic derivation. It has already been pointed out in connection with the discussion of the preoral lobe of the *actinotrocha* that features of the Bryozoa must also be taken into account. For this reason a closer comparison is deferred to the next chapter. But it may be said here that the anterior lobe (with the exception of its ciliary ring) should probably be interpreted as an ancient adult character.

In cases where the anterior lobe is uniformly ciliated and with less of a distinct umbrella-shape (Fig. 9), the larvae generally seem to have a

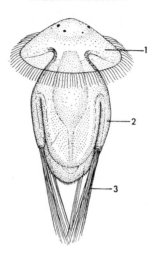

FIG. 8. Larva of *Argyrotheca*. 1, anterior lobe; 2, mantle lobe; 3, setae. (After Kowalevsky, from Korschelt and Heider, 1936.)

short pelagic life and to have undergone major changes connected with lecithotrophy. But there exists obviously no tendency towards such an almost direct development as found in the Ecardines.

After the transition to a sessile mode of life, the larvae of the Testicardines undergo a fairly far-reaching metamorphosis. This consists mainly of the forward inversion of the mantle folds and the development of shells and tentacular apparatus. At the same time the anterior lobe is

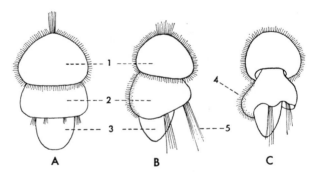

FIG. 9. *Tegulorhynchia nigricans* (Testicardines). Three stages in development of the larva. A, stage with emerging setae, dorsal view. B, almost mature and C, fully mature larva, both seen from the left. 1, anterior lobe; 2, rudiment of the mantle; 3, peduncle; 4, longitudinal ventral ciliary band; 5, setae (After Percival, 1960.)

strongly transformed. According to Percival (1944, 1960) it gives rise in *Terebratella* and *Tegulorhynchia* to the entire anterior portion of the adult body with the tentacular apparatus, etc.; and in addition the mouth is said to be formed on the extreme anterior part of it. According to Conklin (1902) and Plenk (1915) the mouth of other forms is developed behind or exactly at the edge of the anterior lobe. If these different statements are correct, the Testicardines would present great variations in this respect. Originally the anterior lobe seems to have been a purely preoral formation. In *Argyrotheca* it is umbrella-shaped and provided with strong cilia at the edge. There is thus close agreement with the preoral lobe in an *actinotrocha* apart from the fact that the preoral lobe of this larva is not delimited dorsally from the rest of the body.

I wish to point out that the observed similarity between the set of tentacles of the Ecardines larva examined by me (Fig. 7) and that of the *actinotrocha* seems to be no unique occurrence. To judge from an illustration given by Ashworth (1916, Pl. 4, Fig. 5) of a *Lingula* larva the same also applies there, and it might be suspected that the phenomenon is common to all forms belonging to the Ecardines. It is strange that this matter has not attracted any closer attention.

The similarity to an *actinotrocha* is surprising, both because it applies only to the Ecardines (thus not to the Testicardines) and also because of the essential dissimilarities between the adults of Brachiopoda and Phoronida. Could it be merely a similarity resulting from convergent evolution? I do not think so. As far as I can see the situation can be explained by the principle of adultation. (For more detail see p. 54 *et seq.*)

It may be pointed out that the tentacles of the Ecardines larvae are not only organs for floating but also for swimming, and in addition have a function in the ingestion of food. Whether or not their function in the *actinotrocha* goes beyond that of promoting suspension is still not quite clear. As is well known this larva possesses a special organ for swimming in the form of a strong ring of cilia at its posterior end. The fact that the similarity between the tentacular apparatus of the Ecardines larvae and that of the *actinotrocha* is nevertheless so great adds to the importance of this apparatus as an indicator of a comparatively close relationship between Brachiopoda and Phoronida.

It is interesting that in the Ecardines development has become almost direct, but does not entail the simultaneous passage of the juvenile form to the biotope of the adult which is otherwise the rule. This retention of a pelagic mode of life seems to have been made possible by the fact that the special method of feeding found in the adult is also possible during free floating. In either case the food is essentially the same.

It might be mentioned finally that Percival (1960) has established in *Tegulorhynchia* the presence of a longitudinal ventral band of cilia on the middle section of the body which enables the larva to creep about on the substratum (Fig. 9). This is obviously a remainder of the same very early ancient adult character which is found in a number of phyla (see p. 244).

Comparative review of the Tentaculata

In the above treatment of the three groups of tentaculates only exceptionally have comparisons been made between them. It therefore remains to add some comment on this aspect, and to examine to what extent certain characters of the larvae can supply information about the phylogeny of and within these groups.

All three groups of tentaculates are sessile in the adult phase. This condition has influenced the organization not only of the adult, but also of the larva. The characters that have arisen in the adult as a result of the sessile mode of life have in several, but not all cases subsequently been accelerated by adult pressure to the pelagic larva. Furthermore, in the larvae of the tentaculates there also exist characters which in all probability are still older, even dating back to the evolutionary phase prior to the appearance of a sessile mode of life. These very old adult characters are of the greatest importance in any attempt to form an opinion about the phylogeny within the Tentaculata. This is of particular importance since the homogeneity of the phylum has been questioned.

The characters that will first be considered are the adhesive organ (internal sac), the tentacles, the shells, the mantle, and the pyriform organ. We shall examine, on the basis of the principles presented in this book, what can be said about the appearance and disappearance of these organs and parts of the body in the evolutionary tree of the tentaculates (Fig. 11).

It is a common characteristic of all three groups of tentaculates that the adult is fixed in or to the substratum by a part of the ventral side, situated between mouth and anus. In the Bryozoa this part may be more or less extended and flattened, while in the others it forms a tube-like process which is especially long in the Phoronida and the Ecardines among the Brachiopoda. The part in question is already developed in the larvae; in those of the bryozoans and phoronids as a marked invagination of the body wall (internal sac and metasome pouch, respectively); in those of the brachiopods as a process (peduncle). In the bryozoans and phoronids the invagination is everted at metamorphosis leading to a condition which in principle is the same as in the brachiopods. The fact that in the phoronids a loop of the intestine is carried through at eversion and that certain other details are unlike in the three groups cannot prevent the impression of a close agreement with regard to this adhesive organ. For this reason I believe in at least partial homology. This view is held also by Brien and Papyn (1954) at least as far as bryozoans and phoronids are concerned. (They express no opinion about the brachiopods.)

We must imagine that the first phylogenetic appearance of the adhesive organ took place in the free-creeping common ancestor of the three groups (Fig. 10A) that existed ''some time'' before the sessile primeval

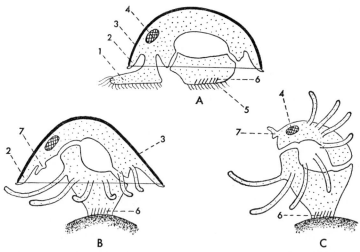

FIG. 10. Diagrams of hypothetical ancestral forms of tentaculates in adult stage. A, free-living form, gliding on the substratum mainly by means of a preoral creeping organ with ciliation on its lower surface (1). B and C illustrate two alternatives for organization of the sedentary primeval tentaculate: B with, C without mantle and shell. 1, creeping organ; 2, mantle lobe; 3, shell; 4, brain; 5, postoral band of cilia; 6, adhesive glands; 7, epistome (as stated in the text it is uncertain whether this structure has to be regarded as a remnant of the creeping organ or of the entire preoral portion of the body).

tentaculate. There is reason to believe that it started as a rather diffuse region with simple adhesive glands, and that the animal was only temporary fixed to the substratum. Gradually, however, the sessile mode of life became permanent, and the body wall with the glands probably became somewhat extended. (At this stage in phylogeny the tentacular apparatus must have been well developed—see below and Fig. 10B and C).

It can be imagined that the first phylogenetic appearance of the adhesive organ took place in the adult phase (cf. p. 221). It was, however, probably "soon" accelerated to the pelagic larva. When the adult became permanently sedentary, the larva must have possessed a simple group of glandular cells for attachment (Fig. 11:2).

The special conditions which now exist in the different groups were differentiated during continued evolution. Thus in the separate evolutionary lines leading to Phoronida and Bryozoa, perhaps even in the line common to these two groups (after the splitting-off of the Brachiopoda), the adhesive organ became inverted. This inversion (not the adhesive organ as such) is obviously a purely larval character.

The essential similarity between the adults of the three groups, as far as the tentacular apparatus is concerned, is so striking that their phylogenetic connection cannot be doubted. In other words, the adult of the primeval tentaculate must already have been in possession of a tentacular apparatus which has subsequently been retained in the different evolutionary lines, even if it has been subjected to certain changes, especially in the line leading to the brachiopods. This view, which has been held for a long time, seems to receive further support from the application of the principles presented in this book (see below).

It is perfectly in agreement with our general experience that the tentacular apparatus is not restricted to the adult, but is also, in certain cases, found in the pelagic larva, viz. in the Phoronida, and the Ecardines among the brachiopods. In the Bryozoa and the Testicardines the larval phase of the life cycle shows, on the other hand, no trace of this adult character. This essential difference between the groups must be due to the fact that in this respect adult pressure has not had an effect on all of them.

My opinion on the evolution of the tentacular apparatus is as follows. The first appearance of this apparatus for feeding took place in the benthic adult phase in the common evolutionary line of the three groups of tentaculates (Fig. 11). We must assume that in the sessile primeval tentaculate the apparatus was fully evolved (Fig. 11:3). The pelagic

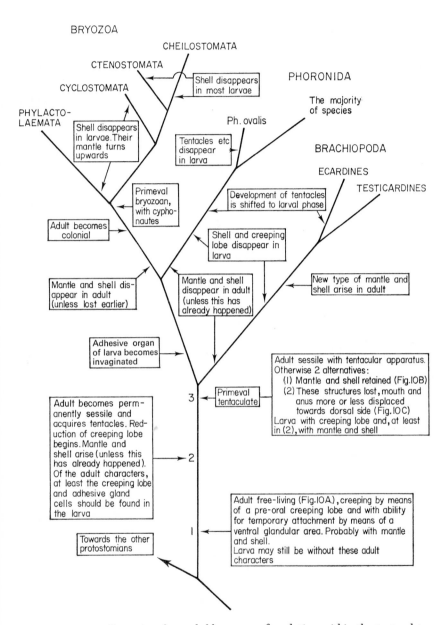

Fig. 11. Diagram illustrating the probable course of evolution within the tentaculates, indicating the stages and changes discussed in the text. Their position is in most cases only approximate.

primary larva of the primeval tentaculate was, however, still devoid of tentacles (cf. p. 220). This condition was retained for some time, even after the separation of the evolutionary lines leading to Brachiopoda, Phoronida, and Bryozoa, and probably also after those leading to Ecardines and Testicardines had become separated (Fig. 11). Subsequently adult pressure caused the development of the tentacles to be transferred to the pelagic larva. This acceleration took place separately in the two special evolutionary lines leading to the Phoronida and to the Ecardines (see the notes in the diagram). The situation produced in this way was then retained up to recent times.

If I imagine the transfer of the development of the tentacles to the larval phase to have occurred at the phylogenetic stages shown in Fig. 11 it is mainly because nothing indicates that at any time tentacles existed in the *cyphonautes* and in the larvae of the Testicardines. The possibility that the absence of tentacles in these two types of larvae is only secondary is diminished by the general rule that adult characters are not lost in the larvae (or at least not in planktotrophic ones) so long as they persist in the adults (see p. 221). It is known, on the contrary, that many larvae stubbornly retain characters that have disappeared in the adult phase long ago.

In spite of what has been said one still feels some doubt about when the development of tentacles was transferred to the larva in the brachiopod branch. The reason for this hesitation is that in *Phoronis ovalis* the development of the tentacles has probably been shifted back to the adult phase (see p. 30). The objection might then be raised that the larvae of Testicardines which likewise are lecithotrophic might have also lost the tentacles. The objection, however, has not much force. Unlike the larvae of the Testicardines the larva of *Ph. ovalis* has undergone an extremely far-reaching morphological transformation and is no longer pelagic. For this reason I consider it most probable that the lack of tentacles in the larvae of the Testicardines is not due to retarded development, but to omitted adultation.

Even if, contrary to what might be supposed, adultation with reference to the tentacles took place somewhere before the splitting of the brachiopod branch into the two subgroups, the phenomenon should at all events have occurred independently in at least two places in the evolution of the tentaculates: both in the brachiopod branch and in the special line leading to the Phoronida. The idea of parallel adultation in the case of the tentacles cannot be avoided unless the evolutionary diagram is modified in such a way that the special lines leading to the Phoronida and the Brachio-poda are made to issue from a common branch, or by the assumption that in the past the *cyphonautes* was provided with tentacles. Neither seems possible.

There does exist, however, a remarkable similarity between the Phoronida and the Ecardines with regard to the tentacular apparatus of the larvae (see p. 47). This might possibly indicate that these two groups were more like each other in this respect, at the time when the develop-ment of this apparatus was passed on to the larva than at present. It appears probable, however, that they already differed considerably in other points.

In the Phoronida, as well as in the Brachiopoda, the larva has at the fore end a more or less umbrella-like structure (preoral lobe, anterior lobe). According to the available information this part of the body is distinctly delimited all around, including the dorsal side, in Testicardines (Fig. 8), while no such dorsal demarkation exists in Ecardines and Phoronida (Figs. 2 and 7). This difference, however, need not exclude the interpre-tation that in all cases we are dealing with one and the same structure, especially since the difference occurs *within* the brachiopod group.

In the Bryozoa nothing corresponding to the above structure seems to exist. I must, however, express my doubts about the correctness of such an interpretation. What might, in the bryozoans, be thought to corre-spond to the preoral lobe is the mantle in a *cyphonautes*. This is a preoral and dorsal differentiation which admittedly envelopes the whole body, on account of its size and therefore presents a rather different appearance, but which on closer investigation can very well be compared with the preoral lobe of the other larvae of tentaculates. The series of diagrams in Fig. 12 should show better than a verbose description that the difference between a *cyphonautes* and an *actinotrocha* is essentially one of proportions only. For this reason I shall restrict myself to a few comments.

I have arrived in three steps at a picture of the *actinotrocha* which is simplified in two respects. It lacks tentacles and the ring of cilia at the posterior end. It seems obvious that for the purposes of comparison we can disregard these structures. As far as the tentacles are concerned, we have already found it probable that the occurrence of this adult character in the *actinotrocha* is due to relatively late adultation. The transference of the tentacles to the larva is a process that *in this connection* is totally irrelevant. It need not necessarily have taken place. (As already stated, everything indicates that it has not occurred in the Testicardines.) The ring of cilia, on the other hand, is a purely larval (adaptive pelagic) character and as such cannot reasonably interfere with the comparison made.

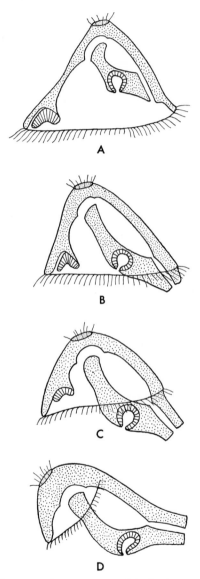

FIG. 12. Diagrams of the essential organization of the body in *cyphonautes* (A) and *actinotrocha* (D). B and C, imaginary intermediate forms intended to show that the two larvae can be compared in spite of their great dissimilarity.

In *cyphonautes* the shells are in intimate contact with the mantle. The latter is the matrix of the shells, and for this reason it is unavoidable to consider the two structures together as a single complex (cf. the analogous situation in the Mollusca). And as I have previously (p. 40) arrived at the conclusion that the shells of *cyphonautes* are an ancient adult character, I am compelled to interpret the mantle as a character of the same kind.

It seems probable that the shells were formed in some ancient adult as dorsal protecting structures, and that the mantle lobes have evolved in connection with the enlargement of the shells in a ventral direction. This enlargement progressed so far that the shells were able to enclose the whole body, as is still the case in *cyphonautes* and also in numerous other instances, as in the molluscs and several groups of crustaceans. I consider it improbable that thin, delicate mantle lobes should have been formed without connection with overlying shells.

I want to stress that this reasoning refers to the adult. When the onto-genetic emergence of mantle and shells was shifted to the pelagic phase of the life cycle, a larva resulted which, although perhaps not identical with a *cyphonautes* in every detail, agreed with it in essential points. (It appears most likely that in addition to shells and mantle this larva also possessed an adhesive organ and a locomotory preoral lobe; see below.) Compara-tively heavy shells seem, however, rather unsuitable for a pelagic larva (cf. above, p. 40). In my opinion they have been eliminated in all tentaculate larvae with the sole exception of *cyphonautes*. The underlying mantle has, on the other hand, been retained everywhere, even if it has been subjected to certain transformations in several cases. In a number of larvae it was transformed into a more or less umbrella-like structure (the preoral lobe) especially suitable for life in the pelagic region (*actinotrocha*, the larvae of certain Testicardines). However, the mantle fold as such must be inter-preted as an ancient adult character, as already pointed out above.

As has just been suggested, the elimination of the shells in the larvae could be imagined to be connected with their unsuitability in a pelagic creature. It must, however, be stressed that after the disappearance of the shells in the adult phase a new adult pressure has come into play, this time a negative one, which has worked in the same direction (cf., e.g., p. 106).

It is tempting to see whether something could be said about where in the evolutionary tree of the tentaculates the mantle and shell were initiated and altered. There is some basis for such a discussion.

Since the mantle (preoral lobe, anterior lobe) is found in the larvae of all three groups, and can be considered as one and the same structure which is (*cyphonautes*) or has been (all the others) the matrix of a shell, and since these structures are ancient adult characters, the logical conclusion is that mantle and shell must already have occurred in the adult phase in the common evolutionary line of the groups, i.e. below point 3 in the evolutionary diagram in Fig. 11. Otherwise we should not be dealing with homologous structures in the three groups.

The common evolutionary line contained two ancestral forms which in certain respects exhibited essential differences: first a free-living (creeping) form and after it the sessile primeval tentaculate, i.e. the last common ancestor of the three groups.

It is not impossible to imagine that the adult of the primeval tentaculate was provided with mantle and shell above the tentacular apparatus. We can well imagine that the tentacles were extensible outside the margin of the mantle in the way shown in Fig. 10B; in addition, features of the brachiopods, especially in a form like *Crania*, must be kept in mind. I find it not only possible, but indeed probable that mantle and shell were already present in the preceding, creeping ancestor (Fig. 10A). In the adult of the primeval tentaculate the structures in question might even have been reduced or altogether eliminated (alternative C in Fig. 10). At present, however, it is not possible to arrive at any more positive conclusion. The only thing for which we have any indication is, as mentioned above, that mantle and shell were evolved in the adult in the common evolutionary line of the three groups. (On p. 41 I have discussed the question of the evolutionary stage at which mantle and shell must have disappeared from the adult in the special line leading to the Bryozoa. See also Fig. 11.)

Because of the uncertainty about the shell and mantle in the adult it is also impossible to assert with any certainty where in the course of evolution the development of these structures has been accelerated to the larva. If they were already eliminated in the primeval tentaculate (alternative C in Fig. 10) we should be forced to conclude that its larva *did possess* both mantle and shell. This might appear paradoxical, but is perfectly in agreement with the principle of adultation (an adult character cannot be shifted to a larva after its disappearance in the adult!). On the same principle it must be maintained that if the primeval tentaculate still possessed mantle and shell it is not certain that these structures were found in its larva. In this case they may have been shifted independently to the larval phase at several stages above point 3 in the evolutionary diagram.

The mantle is retained in the larvae of all three groups, but, with the exception of some few bryozoan larvae (the *cyphonautes*-type), the shells have disappeared. It is impossible to say exactly where in the evolutionary tree they were lost. We only can assert that they disappeared in the larvae later than in the adults, and independently in several instances (see the indications in the diagram, Fig. 11).

We shall now turn to the so-called pyriform organ which as far as is known is restricted exclusively to the larvae of the bryozoans. In the foregoing (p. 44) I have arrived at the conclusion that it is the larval homologue of an ancient organ of locomotion in the adult phase. Since the primeval tentaculate was a sessile animal (nothing else could reasonably be supposed), it is immediately clear that the organ functioned in an ancestor older than the primeval tentaculate. This free-living ancestral form was, in fact, the first notable one in the common evolutionary line (Fig. 11:1). In the adult stage it moved by creeping upon the substratum by means of the preoral portion of the body which was ventrally ciliated. This was probably a retractile creeping lobe, thus basically the same as that still found in *cyphonautes*. In Fig. 10A I have attempted to represent this organ of locomotion in use.*

In connection with the adoption by the adult of a sessile mode of life the organ lost its locomotory function, and for this reason it appears probable that it was somewhat reduced in the primeval tentaculate.

From our general experience (the principle of the influence of adult pressure) we are forced to suppose that the ontogenetic emergence of the creeping organ had already been shifted to the pelagic larva before the initiation of sessile life. It was, in other words, transformed into the pyriform organ. As such it has survived in the recent *cyphonautes* and, although more or less reduced and changed, in the majority of the other bryozoan larvae. In the special evolutionary lines leading to Phoronida and Brachiopoda it has, however, been lost in all larvae without leaving any observable traces.

I have just mentioned that the creeping lobe seems to have been reduced already in the adult primeval tentaculate. It is self-evident that the reduction of this preoral organ continued in the special lines leading to the three groups.

The question arises, however, as to whether or not the creeping lobe

* In Fig. 10A I have also drawn locomotory cilia in a postoral position (5), since I consider that such cilia were found there, too, arranged in a longitudinal band (cf. below p. 244 and Figs. 9B and C).

has completely disappeared in the adults of the recent tentaculates. All of them with the exception of the bryozoans belonging to the Gymnolaemata are known to possess a small preoral structure, the epistome. I dare not exclude that this *might* be a last trace of the creeping lobe.

A homologization of this kind is opposed by the fact that in the *actino-trocha* and the larvae of the brachiopods nothing has been established that might correspond to the pyriform organ of the bryozoan larva. Otherwise it is the rule that an adult character disappears first in the adult and then in the larva. In groups other than the Gymnolaemata the opposite would then be the case. For the present I dare not express an opinion concerning the epistome.

In this connection it must, however, be kept in mind that, if the epistome is in fact a remnant of the creeping lobe, everything leads to the conclusion that the epistome should not be considered a protosoma as is usually assumed, since the creeping lobe cannot very well be interpreted as more than part of this region of the body (see also below).

Even if the preoral creeping lobe can be interpreted as a very ancient locomotory organ, it has evidently had a preceding structure of a different kind but with the same function, viz. a longitudinal postoral band of cilia of the same type as has been established in the larvae of several phyla (see p. 244). As mentioned above (p. 52), Percival (1960) has found just such a ciliated band in a genus of the Testicardines (Fig. 9).

The results at which I have arrived in the discussion of the adult characters of the larvae have some important consequences. First it should be noted that the mantle and shell of recent brachiopods have nothing to do with the structures with the same name in *cyphonautes* and the corresponding structures in other larvae and ancient adults. The mantle lobes and shells of the brachiopods are differentiations which are peculiar to this group and situated in an entirely different region of the body. The approximate position of their evolutionary emergence is marked in the phylogenetic diagram (Fig. 11). It is of interest that recent brachiopod larvae, particularly distinct within the Testicardines (Fig. 8), exhibit both kinds of mantle structures.

In the foregoing treatment of the mantle–shell complex I have pointed to the close morphological agreement with the corresponding structures in Mollusca, as does Lemche (1963). But unlike this author I am unable to interpret the agreement otherwise than merely as a phenomenon of convergence. There is no doubt that the formation of a mantle with superimposed shell has taken place independently in several instances within the metazoans. Here I cannot enter more closely into Lemche's comparisons between *cyphonautes* and Mollusca. My doubts are first of all due to the fundamental difference between the free-living ancestor of the tentaculates and the molluscan type with regard to the locomotory organ of the adult: in the former a preoral creeping lobe, in the latter a postoral one. This taken together with other points which I cannot enter into here, indicates that the relationship between Mollusca and Tentaculata is so remote that mantle and shell of *cyphonautes* can hardly be phylogenetically connected with the structures of the same name in the molluscs.

It is worth mentioning that the occurrence of a preoral creeping organ in the free-living ancestor of the tentaculates also removes this form (and with it the bryozoans!) far from the free-living ancestor of the Entoprocta which, I maintain (p. 108), had a postoral creeping lobe. This is a further argument against the linking-up of Entoprocta and Bryozoa which is still advocated by Marcus (see also p. 115). In this connection it is of great interest that it is the Entoprocta and not the Bryozoa that agree with the Mollusca so far as the creeping organ is concerned (Jägersten, 1964; cf. also p. 247).

If we want to find a structure which in morphology and function tolerably agrees with the preoral creeping lobe of *cyphonautes*, we have to turn to the deuterostomians. I refer first of all to the cephalic shield of the Pterobranchia, especially in *Cephalodiscus*, but doubt whether a true homology applies here. The cephalic shield is, like the "acorn" of the helminthomorphic enteropneusts, the foremost of the three main sections of the body (protosoma). It is, however, impossible to consider the creeping lobe of the ancestral forms of the tentaculates and of the *cyphonautes*-larva as more than a part of the protosoma.

Cyphonautes is the most primitive among the larval types of the tentaculates in the sense that it has retained ancient adult characters to a greater extent than the others. Of the three characters discussed above—mantle, shell, and creeping organ—the other larval forms have retained only a more or less reduced mantle. In other words, in the Phoronida as well as in the Brachiopoda the larvae have undergone a reduction of some ancient adult characters that in some way has reduced the difference from their recent adults. A further reduction of the difference has taken place in those forms in which recent adult characters have been shifted to the larval phase. This applies in a particularly high degree to the Ecardines, where it has resulted in practically direct development.

In the bryozoans, on the other hand, no such reduction of the difference has taken place. The differences between the two phases have on the contrary increased as the result of the appearance of new characters in the adult in connection with the transition to sedentary life and the formation of colonies. This in turn has resulted in the very radical metamorphosis.

Not much needs to be said about the purely larval (pelagic) characters. Of these the most noticeable are the ring of cilia at the edge of the mantle (the "corona" of the bryozoan larvae) and the perianal ring which occurs only in the *actinotrocha*. Both are organs for swimming. The cilia on the edge of the mantle have often been considered homologous with the prototroch of the *trochophora*-larva. Personally I am somewhat doubtful, but do not think that the hypothesis can be rejected out of hand. It probably implies that all protostomians have or have had larvae with a preoral ring of cilia. It should be noted that not only the true *trochophora*-groups but also primitive platyhelminthes and rotifiers exhibit such a ring (see pp. 84 and 68). This does not by any means imply that I agree with the authors who maintain that the larva of the tentaculates are altered true *trochophora*-larvae, i.e. of the type which is found, for example, in annelids and molluscs. A typical *trochophora* is characterized by more than just the preoral ring of cilia.

It perhaps ought to be pointed out that my interpretation of the mantle as an adult character does not directly preclude the view that the larvae of the tentaculates possess a genuine prototroch. It seems possible to me that the formation of the mantle at the time when it was shifted to the larva was effected just below the already existing prototroch, which thus became pushed to its margin—to some degree a parallel to the conditions at the formation of the velum in the larvae of molluscs.

The homologization of the mantle of *cyphonautes* with the preoral lobe in the larvae of phoronids and brachiopods, and the general understanding of the presence or absence of adult characters in larvae have supplied us with quite new possibilities for the critical examination of the tentaculates. We can now see easily how it is possible that these groups, which agree in certain essential features, can possess pelagic larvae with such considerable morphological differences as occur (cf., e.g., Figs. 2A, 4A, 7, and 8).

The differences between the larvae of the tentaculates are essentially due to the following factors:

1. Not all of them have undergone adultation, so far as the tentacular apparatus, their most striking recent adult character, is concerned.
2. Some of them have lost certain important ancient adult characters (shell, creeping lobe).
3. There is a certain progressive divergence in the pelagic character represented by the ciliary covering.
4. They display different kinds of nutrition (planktotrophy versus lecithotrophy).

None of these differences is sufficiently fundamental that a relatively close relationship between the three groups of tentaculates could be questioned. By explaining certain of the previously very puzzling differences the present examination has on the contrary provided further support for such a relationship.

From the above it ought to be evident how important it is that close attention is paid to retained ancient adult characters. Through such characters the tentaculates, perhaps more than any other phylum of the coelomates, enable us to reconstruct the evolutionary processes. This is the reason why I have treated the tentaculates in considerable detail and have even ventured to propose a fairly detailed diagram of their phylogenetic connections. This, like the whole foregoing discussion, must be taken as the *attempt* it actually is, aimed at analysing conditions as thoroughly as possible on the basis of the new principles advanced in this book.

Aschelminthes

It is often maintained that there is no close relationship between the different groups, Gastrotricha, Rotatoria, Kinorhyncha, Priapulida, Nematoda, Nematomorpha, and Acanthocephala, which are usually united under the heading Aschelminthes. Admittedly they are very dissimilar in a number of cases, and the existing similarities are to a large extent negative. In the respects that interest us here they are, however, in close agreement, yet from the taxonomic point of view this agreement is perhaps not very elucidative (see below).

All these groups can be said to have a direct development. In maintaining this point it seems that I must be compelled to admit quite a number of exceptions. The "larvae" which can be distinguished in many forms are, however, not primary larvae. It is common to all groups that the original life cycle with a pelagic primary larva and a benthic adult has been modified to a cycle with direct development. Yet subsequently new changes have taken place in several groups and have led to secondary larvae. As these conditions are in part of a very special nature, I shall discuss each group by itself. About the majority of them, however, not much is to be said.

Gastrotricha

Here conditions seem to be fairly simple. From the egg, which is surrounded by a shell-like envelope, hatches a juvenile which resembles the adult. According to de Beauchamp (1929) the holoblastic cleavage results in a coeloblastula. Thus, in spite of many other changes, an original feature is retained in the embryogenesis (cf. p. 242).

Rotatoria

As is usual in cycles with direct development the newly hatched/ new-born juvenile of most rotifers resembles the adult. Exceptions are found in some sedentary forms like *Cupelopagis*, and other genera within the group Collothecacea. In some of these forms the adult (female) is provided at the anterior end with a crown of processes which resemble tentacles. In this and other features there is a considerable dissimilarity with their juveniles which, generally speaking, are more like the adults of the other monogonate rotifers (Fig. 13). The dissimilarity between juvenile and adult of the sedentary forms is due to the fact that the adult has been changed in consequence of its special mode of life. Because of the divergence between the two stages the less altered, free-swimming juvenile is occasionally called "larva". This may be justifiable, so long as it is kept in mind that this is a kind of secondary larva (see p. 8).

The sometimes rather pronounced similarity between the free-swimming forms and a *trochophora* has given rise to the theory that the rotifers must be derived from neotenic annelid larvae. This theory has, however, hardly any adherents today. This is due mainly to the investigations of de Beauchamp (1907, 1909) who finds the most primitive organization not in the pelagic but in the benthic rotifers. In the latter the ciliary covering (the "corona") is used for creeping rather than free-swimming, and it is probable that the ancestral forms were provided with a ventral ciliary covering similar to that of the primitive *Gastrotricha*. The morphology of the corona in the pelagic rotifers must therefore be a result of adaptation to locomotion by swimming and to a new method of feeding.

It is, however, striking how similar the food-ingesting ciliary apparatus of certain rotifers is to that in those *trochophora*-larvae which I have found

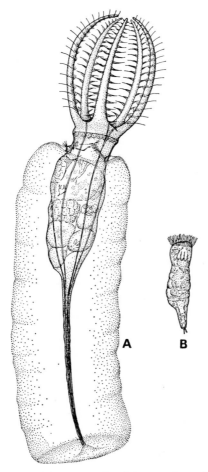

FIG. 13. *Stephanoceros fimbriatus* ♀. A, sessile adult animal; B, free-swimming, newly hatched young. Here we have an example of considerable dissimilarity between the juvenile and the adult, arisen in connection with the adoption by the adult of sessile mode of life. Because of this dissimilarity the juvenile can be designated as a secondary larva. (After Jurczyk, 1927—simplified.)

to be the most original in this respect (see pp. 108, 121, and 159). Just as
in these larvae, there exists between the preoral locomotory ring of cilia
(trochus) and the postoral one (cingulum) an *adoral* zone with finer cilia
in which, as indicated by the name, food particles are transported to the
mouth (Fig. 14).

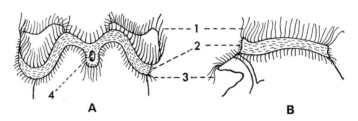

Fig. 14. Anterior portion of the body of *Pedalia*. A, ventral view; B, seen from the left.
(After de Beauchamp, 1907.) 1, trochus; 2, adoral ciliary zone; 3, cingulum; 4, mouth.
(Cf. Fig. 44.)

According to Wesenberg-Lund (Kükenthal-Krumbach, "Handb. d.
Zool., Rotatoria", p. 19) this type of feeding apparatus is found e.g. in
the pelagic genus *Pedalia* (Fig. 14) and in the benthic family Flosculariidae
(Melicertidae).

How is this agreement with an original *trochophora* to be understood ?
There are only two possibilities. Either we are faced by a phylogenetically
conditioned similarity, thus a similarity which can be traced back to conditions in the common ancestral form of both rotifers and *trochophora*
groups, or else this is nothing more than a phenomenon of convergence.
According to the prevailing opinion the latter is the case. At present in-
sufficient experience with the rotifers prevents me from expressing a
definite idea in the matter. However, I want to stress that the view, in all
probability correct, according to which the ventrally extended ciliary
apparatus of certain benthic forms is original, does not necessarily imply
that a ciliary covering consisting of rings of cilia resembling those in a
trochophora must be derived from the former apparatus. In other words, it
is possible to combine de Beauchamp's idea of what is original with the
view that the mentioned similarity with a *trochophora* can be traced back
to the conditions in a common ancestral form. I shall briefly develop this
seemingly incongruous way of reasoning.

According to the basic idea underlying this book the life cycle of the
ancestors of the rotifers consisted of a pelagic larval, and a benthic adult

phase, provided we go sufficiently far backwards in evolution. It is conceivable that the primary larva of the pelago-benthic ancestor had the ciliary apparatus with adoral zone characteristic for the original *trochophora* type (cf. Fig. 44, p. 160). In this respect at least it was a *trochophora*. The adult, on the other hand, possessed a ventral ciliary covering adapted to creeping locomotion (cf. p. 244) and took up its food in another way. In one or several of the evolutionary lines leading to the recent rotifers the usual change to direct development took place. In other words, the pelagic primary larva was lost, with holobenthic life as the result. This is the situation that we find retained in those rotifers, which according to current opinion are primitive.

In one or more other lines within the evolution of the rotifers the pelago-benthic life cycle was transformed into holopelagic life, and this change was effected in a neotenic way. (About such a course of evolution see p. 233.)

In this way the larval ciliary apparatus and the larval method of feeding became characteristics of the new (secondary), pelagic adult. It must be observed that even in this case the result was direct development, although in an entirely different way (see p. 236).

It might be added that nothing prevents the ever present adult pressure from having caused a number of adult characters to be shifted to the primary larva *prior* to this change to a holopelagic life cycle. Such changes would have mitigated the differences from the holobenthic rotifers discussed above. It is certain, on the other hand, that later introduced changes in a number of different directions have led to the greatly diversiform morphology with regard to details which characterize the rotifer group. Conditions may also have become complicated by a later shift from benthic to pelagic life and vice versa. This might explain why certain benthic forms, too, possess an adoral zone (Flosculariidae).

It is now implied that the ciliary apparatus may be of a primitive nature both in benthic and in pelagic rotifers. The two different types may have originated quite independently and in different phases of the original pelago-benthic cycle, one in the adult phase as an adaptation to the benthic region, the other in the primary larva as an adaptation to the pelagic region. If this is true, both types of apparatus are older than the rotifers as such.

According to my attempt at interpretation, a ciliary apparatus for feeding provided with an adoral zone is an original larval feature, wherever it occurs within the group, but *this by no means implies that the forms in question must be derived from neotenic annelid larvae.*

Sedentary rotifers with an adoral zone are closely similar to the *Entoprocta* in so far as the same, originally larval feeding apparatus is found in the adult in either case (cf. p. 108). Its appearance in the adult has, however, taken place independently in the two groups.

I want to stress that the above interpretation is only tentative, and I am thus not expressing any final opinion in the question. The discussion

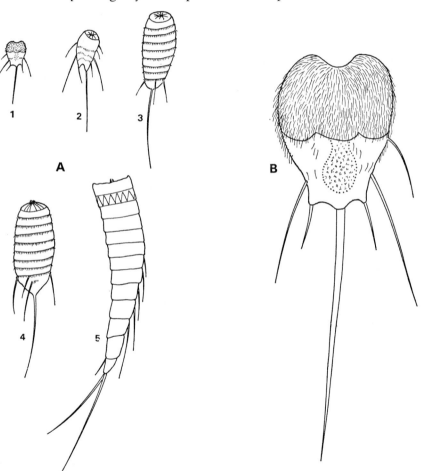

FIG. 15. Diagram representing the post-embryonic development of *Echinoderella elongata*. 1, youngest larval stage; 5, adult stage (anterior end retracted); B, youngest stage under higher magnification. (After Nyholm, 1947.)

demonstrates, however, the importance of keeping in mind both phases of the original life cycle when dealing with phylogenetic questions. For the elucidation of the conditions in the rotifers further investigations are required.

Kinorhyncha

Unfortunately not much is known about the ontogeny of this group. Nyholm (1947) has found a very young stage the appearance of which is shown in Fig. 15B. The anterior part of the body is covered with fine hairs. Their nature is not mentioned; they are, however, evidently not cilia but differentiations of the cuticula. This youngest stage encountered differs very considerably from the adult in spite of the fact that the entire life cycle takes place in one and the same biotope. No metamorphosis in the proper meaning of the term, however, takes place, but the "larva" changes gradually into the adult via a series of intermediate forms that are separated by ecdysis.

It is obvious that this development is considerably altered. As far as is known, no traces of the original pelagic primary larva are retained, neither is it likely that future investigations will reveal anything pointing in this direction.

Thus the ontogeny of these animals must be considered as a direct development in which, however, a divergence between the youngest post-embryonic stage and the adult has taken place, leading, one might say, to a secondary larva. It is open to question, to what extent the few forms examined so far are representative of the group, and until we have more information we cannot say with certainty whether the divergence is due to a change of the juvenile form or of the adult. For this reason further investigations would be highly desirable.

Priapulida

When the fundamental ideas laid down in this book are applied, the priapulids offer interesting information. In spite of the changes in the rest of their ontogeny the early embryonic development contains some primitive features: external fertilization, holoblastic radial cleavage, coeloblastula, and gastrulation by invagination (cf. p. 239 et seq.).

According to Lang (1953) the young reared larva of Priapulus has undergone very little differentiation when it leaves the protection of the egg.

It is without cilia and shell structures, without mouth and anus, and no internal organs are developed. It is thus more or less an embryo. In spite of this the body is strongly contractile, obviously enabling a certain amount of locomotion.

It is known that the larvae so far encountered on the biotope are provided with an armour of cuticular plates (Fig. 16B). In *Halicryptus*

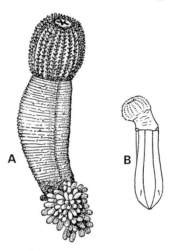

FIG. 16. *Priapulus caudatus*. A, adult, natural size. (From Korschelt and Heider, 1936.) B, larva, magnification *ca.* × 20. (After Lang, 1948.) This is an example of a secondary larva, probably arisen by alteration of the adult.

Purasjoki (1944) has found the smallest of these armoured larvae to be almost exactly the same size as the egg. From this observation he draws the conclusion that the armour is already formed when the larva leaves the protection of the egg. Lang has been able to establish that at the time of hatching *Priapulus* has no armour, but within two days a thin hyaline layer can already be seen over the cells of the ectoderm. He believes this layer to represent the beginning of the armour, which thus is apparently formed at a very early stage.

If we disregard the armour and, as far as *Priapulus* is concerned, the small "foot" at the posterior end (see below) there is on the whole close agreement between larva and adult. In my opinion there can be no doubt that there has been a change from the original type of ontogeny with a pelagic primary larva to holobenthic direct development and that afterwards a secondary divergence between juvenile form and adult has taken place, resulting in the present secondary larva.

The question now is how this divergence has arisen. I think it most probable that it is mainly a result of alteration of the adult. This is indicated by the fact that the larva more than the adult exhibits a similarity to other groups of the aschelminths. In the larva of *Priapulus* we find at least three features to which this applies; these are the cuticular plates, the small "foot" at the posterior end, and certain adhesive tubes.

It has to be admitted that no plates of a similar shape occur in any other group. However, distinctly delimited cuticular structures are found not only in Kinorhyncha but also in certain rotifers, gastrotrichs, and nematodes, and a strong general cuticula exists in the majority of the other aschelminths.

The "foot" resembles to a high degree that in the rotifers, and for this reason it is tempting to consider these structures as homologous, as Hyman (1951b) has done. However, considering the great differences between Priapulida and Rotatoria in other respects and being aware of the number of independently evolved organs for attachment within the aschelminths I find it most improbable that these two features have a common origin. In this final comparison between Priapulida and the other groups of the aschelminths Lang (1953) who had discovered the foot in the larva of *Priapulus* makes no mention of it. This must be interpreted as indicating that he does not consider it as pointing to a closer relationship with the Rotatoria.

Apart from the foot of the rotifers, differently formed organs for attachment are found in several free-living groups. In the Gastrotricha especially, many adhesive tubes are found on different parts of the body. The larva of *Priapulus* possesses two pairs of processes situated laterally at some distance from the posterior end. These are supposed by Lang (1948) to be tactile sense organs. Lang denies categorically that they could have any function for attachment, since they are not connected with any glands. By carrying out experiments I have, however, been able to prove clearly that larvae kept in dishes are able to attach themselves to the glass by means of these processes, often so efficiently that they are not dislodged by the water currents produced with a pipette. This shows that we are after all concerned with some kind of adhesive tube, but as far as I can see no homologization with any particular processes in other groups is possible. (It might be mentioned that according to Hammarsten (1915) adhesive tubes resembling those in Gastrotricha are found at the anterior end of *Halicryptus*.)

Even if none of these three characters of the larva, which distinguish it so markedly from the adult, can be said to have a direct homologue in any other group of the aschelminths, there is still no doubt that it is the larva and not the adult which on the whole agrees most closely with the adults of the other groups. This points distinctly to the correctness of my conjecture that a change, first of all in the adult, has taken place in the Priapulida after the introduction of direct development.

In other words, I have arrived at the conclusion that the larva exhibits at least three ancient adult characters which allow us to form some idea about the ancestral form of the group, and which in this way strengthens the concept that the Priapulida belong to the Aschelminths.

It is possible to form a tentative opinion as to why the adult has been subjected to the above changes. The recent adult lives by burrowing in loose substrates. Where no appendages exist a burrowing mode of life is most suitable for a soft and flexible body with peristaltic ability, and no adhesive organ is required. For this reason it seems reasonable to suppose that the adult has lost the characters discussed above as a consequence of the transition to its present mode of life.

The burrowing mode of life has, on the other hand, led to the acquisition of another character: the essentially radial symmetry of the body. (It is a general rule that such a mode of life, as well as sedentary and pelagic life, produces this effect as a result of uniform contact with the surrounding medium.) The radial symmetry is particularly pronounced at the anterior end (the "proboscis"), where it is conditioned by the terminal position of the mouth and the arrangement of spines and papillae; in this region radial symmetry is complete, as far as the external appearance is concerned.

The anterior end of the larva presents the same situation, but because of the armour the major part of the body is bisymmetric (biradial). Thus the plates—and in *Priapulus* also the adhesive processes—are dorsally and ventrally identical in size, shape, and position. Using material from Gullmar fiord (west coast of Sweden) I have studied these features in detail without being able to observe the slightest difference between the opposite sides.

In the foregoing discussion I have arrived at the conclusion that the larva is more primitive than the adult, *inter alia* with regard to the occurrence of the cuticular plates. On account of the fact that the bisymmetry is conditioned by these plates it can be supposed that this type of symmetry is also an ancient adult character. Thus the course of evolution of the adult seems to have been: bilateral symmetry $\rightarrow$ bisymmetry $\rightarrow$ radial symmetry. The correctness of this view is supported by the fact that a similar sequence can be traced elsewhere in the animal kingdom (Jägersten, 1955, p. 331; see also below, p. 101).

A question which might finally be raised is why the larva, living in the same biotope as the adult, has not been altered throughout in the same way as the adult. By way of an answer I can, unfortunately, only point to the general principle that larvae and juveniles are usually more conservative than the adults. I consider it probable, however, that, speaking phylogenetically, the larval characters are in the process of being suppressed as a result of recent adult pressure. It is remarkable that the larva of *Halicryptus* lacks both the "foot" and the lateral adhesive tubes. If the plates were to disappear, the agreement with the adult would be practically complete, and we should be able to speak of secondary direct development.

Nematoda

Here the ontogeny is strongly altered. The juvenile that leaves the egg as a rule greatly resembles the adult, so that typical direct development is applicable here. The specialists of the group nevertheless often speak misleadingly about "larvae". As far as any more marked difference between young and adult exists, it should only be considered as secondary divergence. I cannot enter into a closer examination of this group with its wealth of species—I leave this to the specialists—but I do consider it possible that such an analysis might reveal interesting aspects.

Nematomorpha

In spite of the great changes within this group as the result of the parasitic mode of life its ontogeny nevertheless exhibits at least two original traits: coeloblastula and gastrulation by invagination (Fig. 17).

As might be expected, no trace of the primary larva is left. This has been replaced by direct development, but in the same way as in the Priapulida a subsequent divergence betwen juvenile phase and adult phase has taken place, with the result that a distinct secondary larva can be distinguished (Fig. 17H).

Our knowledge about this larva is essentially restricted to Gordioidea. The ontogeny of the Nectonematoidea with the single genus *Nectonema* is practically unknown. Through the observations of Huus (1931) we do,

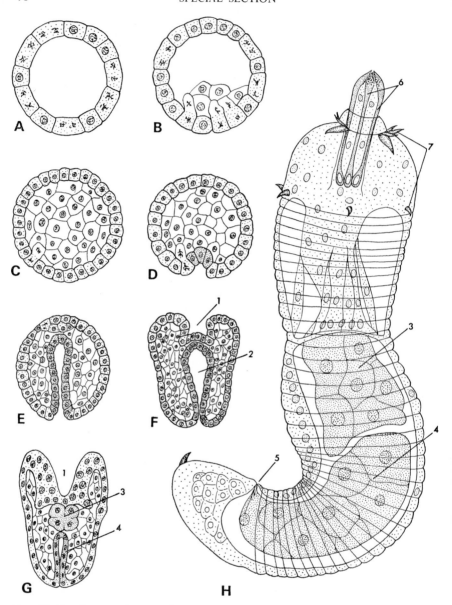

Fig. 17. Developmental stages of *Gordius aquaticus*. A, coeloblastula; B, initial development of mesoderm; C, blastocoel entirely filled with mesoderm; D, initiation of gastrulation; E, gastrulation terminated; F, invagination of proboscis begins; G, poison gland set off from the entoderm; H, mature larva under higher magnification; 1, invagination of proboscis; 2, entoderm; 3, poison gland; 4, mesenteron; 5, anus; 6, stilettoes; 7, spines. (After Mühldorf, 1914.)

however, know that a larva of the same type also occurs in these marine forms (Fig. 18).

A special distinguishing feature of the larva is its poison apparatus. This consists of a large gland with efferent duct opening terminally on a protrudable proboscis at the anterior end, and on this proboscis, around the distal part of the duct, three long stilettoes. The anterior end also possesses radially arranged spines which give the larva a certain likeness to the adults of some other groups within the aschelminths, particularly the Priapulida.

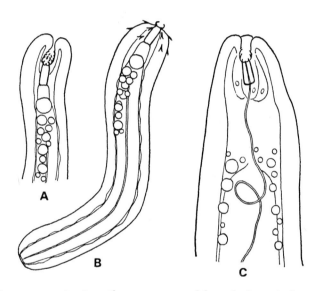

Fig. 18. *Nectonema munidae*. Juvenile stages extracted from the host. A, larva (secondary larva) with inverted anterior end; B, with extroverted anterior end; C, anterior part of a somewhat older stage. (After Huus, 1931.)

The question is now, however, whether or not this similarity is phylogenetically conditioned. This can only be considered possible if the end with the stilettoes is really the morphologically anterior end. This is denied by Mühldorf (1914) who is of the opinion that the end with the stilettoes is only the physiological anterior end.

Mühldorf bases this idea on the earlier embryology. It is known that the blastopore formed at invagination develops directly into what is usually considered as the anus of the larva, and this does not agree with what might be expected in a group within the Protostomia.

A detailed discussion of this question is outside the scope of this book. Certain points of view must, however, be expressed, and I shall say immediately that I cannot share Mühldorf's view. In the light of our present knowledge about the embryogenesis of mouth and anus we are compelled to believe that considerable changes have taken place in this respect in the nematomorphs. This is also indicated by the fact that different scientists have arrived at contradictory results over several questions connected with the early embryology of the different forms. I therefore see no reason for relinquishing the interpretation that the end with the stilettoes is homologous with the anterior end of the other aschelminths. A homologization of this kind is supported not only by the occurrence of the rings of spines but also by the fact that the whole portion of the body upon which they are found can be retracted in the same way as in Priapulida and Kinorhyncha.

It is difficult to tell whether or not this peculiarity in early embryogenesis is connected with the adaptation to parasitism that is implied in the formation of the poison apparatus, but I consider it by no means improbable.

It is probable that the opening which in *Gordius* is called the blastopore and which becomes the anus, is only a small posterior remainder of the original blastopore. We are obviously concerned here with a phylogenetic change of the same kind as in some deuterostomians (see p. 246). In forming a decision as to which should be the anterior end of the body, no more importance can be attributed to peculiarities of the blastopore in *Gordius* than to the fact that in certain deuterostomians with altered ontogeny the blastopore can become the mouth (see Fig. 51).

In the adults of certain genera there is a mouth opening at the anterior end, and it is certain that at one time the larva, too, possessed a mouth. The question then arises whether or not it has disappeared without leaving any trace whatsoever. I do not believe this to be the case.

In the majority of the groups of free-living aschelminths the mouth is situated terminally at the anterior end. In this position the duct from the poison gland in the larva of the nematomorphs also opens. This duct is formed early in embryogenesis (immediately after gastrulation) as an invagination of the ectoderm exactly opposite to the blastopore, and it is then connected with the poison gland. This latter arises from the entoderm by detachment of the inner, blind end (Fig. 17G).

This and other observations suggest that it is the anterior part of the mesenteron that has been detached from the posterior portion and has been transformed into the poison gland. The poison duct is obviously homologous with the ectodermal stomodaeum in the other groups.

Such a transformation could take place without obstacle, once the alimentary canal, which is no longer used for food uptake at any stage, had lost its original function.

It is easy to imagine that the transformation was initiated in such a way that the future entoparasite discharged its digestive and perhaps poisonous mesenteron fluid into the wound produced in the host by the stilettoes. Parasitism probably began as external parasitism merely by sucking-out. Subsequently the worm passed through the wound and became an ento-parasite. Once evolution had reached this stage some other changes be-came possible. The poison apparatus was no longer required in the adult and thus became reduced. In the young it was, on the other hand, still needed and retained, although in an altered condition.

This line of thought obviously presupposes that the infection of the host is, at least originally, due to an action on the part of the larva. If the recent larvae should be ingested with the food—information about this way of infection exists, although the facts are still not completely eluci-dated—it is possible that the poison apparatus would no longer be needed even in the larval stage. The fact that it is nevertheless retained can in this case be interpreted as an example of the phenomenon that juvenile forms are more conservative of characters and organs that are in the process of reduction, i.e. they often exhibit ancient adult characters (see p. 7).

According to this attempt at interpretation of phylogeny there should have been a time when the adult of the nematomorphs was more like the recent larva, especially at the anterior end. In other words, there should have been a time when development was more direct. This idea receives further support from the fact that in its spines and stilettoes the larvae ex-hibits a certain resemblance to the adults of other groups in the aschel-minths.

A marked difference has in any case arisen between the earlier and later phases of the life cycle, so that a secondary larva and secondary metamor-phosis have evolved. This metamorphosis implies that considerable parts of the anterior end of the larval body (how much of it is not fully estab-lished) are shed inside the host. Thus we have an interesting parallel be-tween the metamorphosis in the nematomorphs and that in a number of marine groups and forms with primary larvae. In both cases a shedding of parts of the larval body occurs as a consequence of evolution with strong divergence within the life cycle. The difference does exist, however, that in the nematomorphs it was mainly the transformation of the adult that was responsible for the divergence.

The shedding of the anterior end of the larva does not justify the belief that this is the *cause* of the altered state of the adult in this part of the body. The process should not be interpreted to mean that the individual is "decapitated" in every generation. The loss of the anterior end of the larva and the state of that of the adult are, so to speak, two aspects of the same thing—a final result of an evolutionary transformation by which juvenile form and adult have diverged to a very high degree. It is this divergence that must be bridged in some way at metamorphosis, and it occurs here as in many other animals by the shedding of parts of the body. It should be kept in mind that this ontogenetic process is based on a long process of evolution. In ontogeny fast, radical changes are possible. Phylogeny on the other hand is a slow, gradual process.

As an important result of the analysis of peculiarities within the Nematomorpha it can be pointed out that I, in contrast to Mühldorf, consider the organization of the larva to be of considerable phylogenetic interest. If my interpretation of the situation can be considered as probable, it would supply us with a theoretical basis for future investigations. These should be directed first of all towards ancient adult characters found in the larva, but also towards solving a number of problems connected with early embryogenesis. According to the information and illustrations supplied by Mühldorf the embryogenesis of *Gordius* seems to be altered from what is usual within the Protostomia in the following three respects:

1. The mesoderm is formed in the blastula stage. (For this reason it is difficult to say to what extent it is ento- or ectodermal.)
2. A small remnant of the original blastopore gives rise to the anus.
3. The depression developing into what I interpret as the changed stomodaeum takes place at the pole directly opposite the present blastopore. Originally it probably took place in a more ventral position.

Acanthocephala

The life cycle of this parasitic group is so profoundly altered that no free-living developmental stages are retained. The "mature" embryo ("acanthor") enclosed inside the egg membranes possesses at its anterior end provisional hooks which after its transfer to the intermediate host are lost, together with the embryonic epidermis and other spiny structures situated on it. It may be supposed that these hooks and spines are remnants of a time when the ancestral forms led a non-parasitic life. This might possibly also apply to the hooks on the proboscis of the adult. This possibility is not excluded by the fact that these hooks now serve to anchor the parasite. I thus believe in the correctness of Lang's (1953) statement that there might be homology between the proboscis of the acanthocephalids and of the priapulids.

Platyhelminthes

Among the three groups, Turbellaria, Trematoda, and Cestoda, of the platyhelminths only the Turbellaria exhibit a more primitive life-cycle, and this in the Polycladida which is the only group of turbellarians with unmistakably pelagic primary larvae. These show a certain variation, but most of them can be included under the general denomination "Müller's larva".

It is best to state right away that a more detailed treatment of the complicated and very varying life cycles in Trematoda and Cestoda is out of the question here. I shall restrict myself to a few remarks. First of all it should be stressed that in these groups the parasitic mode of life has caused considerable alterations. It is tempting to believe that the changes in the juvenile phase are entirely a result of parasitism. This, however, cannot be considered as possible. The replacement of the primary larva by direct development is in all probability a fundamental change that has taken place prior to the transition to parasitic life. Because the groups in question are probably derived from forms resembling turbellarians, it may be justifiable to say that the disappearance of the primary larva took place at the turbellarian stage, probably independently in several evolutionary lines. The numerous different larval forms that now exist are all "secondary larvae". I want to stress that this certainly applies also to those ciliated larvae which for a short time are free-swimming, such as *miracidium* in digenic trematodes and *coracidium* in, for example, *Dibothriocephalus*. These result from new divergences between the different phases of the life cycle and are thus certainly not directly derived from the original primary larva (see below). This is separated from the present complicated conditions by an evolutionary period with "ordinary' direct development.

I do not mean to exclude the possibility that the ciliary covering and the free-swimming locomotion of the secondary larvae described is an inheritance from the original primary larva, but because the adult of the turbellarian ancestral forms was ciliated and probably more or less free-swimming in the open water it is impossible to draw sharp limits between the different phases of the life cycle when considering the ciliary covering.

It may be noted that the *cercariae*, which must also be interpreted as a kind of secondary larvae (although differentiated in a later phase of the life cycle), are, in contrast to the younger stages mentioned before, devoid of a ciliary covering. In the *cercariae*, the problem of locomotion has been solved by the appearance of the very efficient tail which, often together with diverse hairy processes and fins that facilitate floating and swimming, constitutes the essential difference from the adult. It is remarkable that an entirely new swimming organ has arisen in this case, instead of a reconstruction of the original conditions with cilia.

The attempt to elucidate further the life cycle of trematodes and cestodes falls outside the scope of this book, and for this reason I now turn to the Turbellaria and first of all to the Polycladida. For a long time the eight lobes have been considered as a characteristic feature of the Müller's larva, and it is the number eight which has led Lang and other scientists to comparisons with Ctenophora and to different hypotheses about a phylogentic connection between this group and the Turbellaria. In this connection it is interesting to note that larvae are found both with fewer and with more numerous lobes. Larvae with ten lobes have been described by Dawydoff (1940a) from Indo-Chinese waters, and I have myself studied such larvae from southern India (Fig. 19B–D).*

These larvae with ten lobes possess above the usual eight two additional lobes near to the single one on the dorsal side. The majority of the lobes are generally quite long, often almost resembling arms. This renders the external appearance of the body rather like that of an *actinotrocha*. This similarity is especially striking in a larva described by Dawydoff, the reason being the exceptionally large unpaired ventral lobe which is reminiscent of the large preoral lobe of the *Phoronis*-larva (Fig. 19D). There is, however, no phylogenetic connection with this larva (see p. 88).

---

* My material of larvae with ten lobes consisted of about one dozen specimens collected on different occasions in February 1964 by netting near the shore at Mandapam Camp opposite Ceylon.

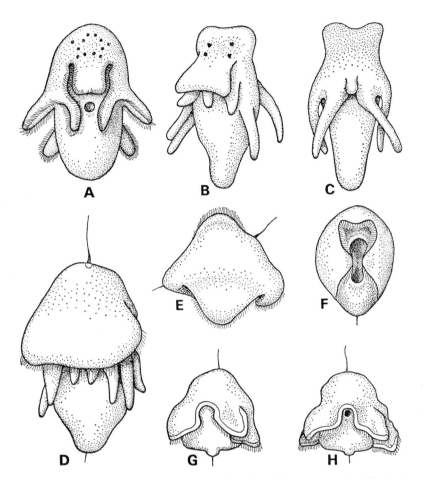

Fɪɢ. 19. Different larval types occurring in the Polycladida. A, typical Müller's larva with eight lobes, ventral view. Only the cilia in the preoral band have been drawn. (After J. Müller, somewhat modified.) B and C, larva with ten lobes, ventrolateral (B) and dorsal (C) view. D, type with ten lobes, preoral lobe greatly enlarged, superficially resembling that of an *actinotrocha*. E and F, Goette's larva, lateral (E) and ventral (F) view. G and H, divergent form with four lobes, almost dorsal (G) and ventral (H) view. (B–H after Dawydoff, 1940a.)

The larvae with ten lobes are mostly bigger, sometimes much bigger than those with eight lobes. Dawydoff has found some real giants with a body length of between 4 and 5 mm. This size leads to the conclusion that these larvae are planktotrophic. Furthermore, in a larva of hardly 1 mm in length I have been able to detect in the lumen of the intestine particles which evidently were ingested food. It can thus be considered established that *at least some Müller's larvae are planktotrophic with a long pelagic phase.* This is an important general statement (see p. 224).

Dawydoff thinks that the larvae with eight lobes might, in certain cases at least, acquire two additional lobes prior to metamorphosis. It is, however, certain that some of them never develop more than eight lobes.

It cannot be decided on present knowledge which number of lobes, either eight or ten, is the original one, neither can we tell whether or not the so-called "Goette's larva" (Figs. 19E and F) is a simplified Müller's larva. It is found by Kato (1940) that Goette's larva, in contrast to Lang's suggestion (1884) is not, or at least not in all cases, an earlier ontogenetic stage of some ordinary Müller's larva. Kato established that in two Japanese species of *Stylochus* it metamorphoses into the adult without a previous increase in the number of lobes. In this connection it is of importance that Kato finds the embryogenesis of Goette's larva to be very similar to that of the Müller's larva.

The situation within the polyclads is further complicated by Dawydoff's discovery of a still more divergent larva, a larva with four lobes, but with an arrangement quite unlike that found in the Goette's larva, consisting of one pair of ventral and another pair of almost lateral lobes (Figs. 19G and H). The ontogenetic and phylogenetic relation of this type to the others is not clear.

Dawydoff proposes the name "*lobophora*" for larvae with a long pelagic phase and with elongated lobes, irrespective of whether their number is eight or ten. I consider the introduction of a new term rather superfluous, at least until the situation is better elucidated. For the present time it might be more justifiable to call all polyclad larvae with more than four lobes, irrespective of their length, Müller's larvae. Variations can easily be distinguished, when the necessity arises.

Ever since Müller's original description it has been known that the ciliary covering of Müller's larva is differentiated into a general, fairly fine ciliation extending over the whole or at least the major part of the body, and a band of somewhat stronger cilia on the lobes, sometimes even forming a connection between them (Fig. 19A). This band is thus in reality a preoral ring of cilia stretching in bights upon the lobes. Because there is a certain degree of agreement in the embryology (spiral cleavage, primitive mesoderm cells, etc.) between the Müller's larva and a *trochophora* this ring of cilia must be interpreted as a prototroch. One could go so far as to say that Müller's larva might be a primitive, but at the same time partially altered *trochophora*—altered especially by the emergence of lobes facilitating floating. The extension of the zone of the prototroch into lobes is not unique. The velum of the gastropods and others provides some kind of parallel (cf. also p. 161—the *rostraria*-larva).

It is remarkable that a larva provided with lobes is not of universal occurrence in the Polycladida. Such a larva has admittedly been established both in *Cotylea*—here in the majority of (or all ?) examined species —and in *Acotylea*, but in quite a number of species within *Acotylea* it is replaced by a larva without any processes whatsoever and with uniform ciliation covering the whole surface of the body. This is a case of more direct development, and this is the term usually applied, although the expression ''larva'' is often retained. This expression can be considered as partially justified, since at least in some cases (e.g. *Notoplana humilis*, the larva of which according to Kato is strongly photo-positive) the life cycle is still divided into a pelagic juvenile and a benthic adult phase.

On the basis of present knowledge we can definitely assert that at least some of those forms within *Acotylea* which display direct development must have earlier had (or, more correctly, their ancestral forms had) a Müller's larva. This can be concluded especially from the genus *Planocera*. According to Kato (1940), *P. multitentaculata* possesses a free Müller's larva, while the embryogenesis of *P. reticulata* includes a stage with well-differentiated lobes which are lost before hatching, so that in this case the Müller's larva is enclosed in the egg (the fact that it is somewhat different, mainly by not having the unpaired dorsal lobe, is of no importance in this connection).

Features in the genera *Stylochus* and *Haploplana* point in the same direction. In the former, according to Kato, five species have a development via a Goette's larva, while two have direct development. It is worthy of comment that Lang (1884) established with some surprise that the two closely-related species *Stylochus pilidium* and *S. neapolitanus* develop quite differently, the former having a Goette's larva, the latter direct development. As, furthermore, *S. luteus* has a typical Müller's larva with eight lobes (see Lang, 1884, p. 403), the ontogeny within the genus *Stylochus* exhibits very varying features. Finally, it may be noted that in *Haploplana villosa* Kato has established direct development, while in *H. inquilina* Surface (1907) found a typical Müller's larva. In spite of this difference Kato established a close agreement between these species in their earlier phases of embryogenesis.

It is obvious that these examples point to relatively recent changes in ontogeny, and observations on the genus *Planocera* in particular lead to the conclusion that evolution has progressed from larva to direct development. The fact that in several cases the species within one and the same genus behave in different ways shows that the elimination of the larva has taken place independently in several different instances.

Even if it can be considered as demonstrated that in several cases, forms which now have direct development at one time had a Müller's larva, we still cannot assume that this larva is original for all polyclads. Considering this question by itself, we might imagine that there are among *Acotylea* forms in which the direct development that occurs is original. In such a case it would, however, be a question of direct development of a different type from that discussed so far in this book. The evolutionary lines leading to these forms should never have exhibited any major divergence between young and adult (cf. p. 91—nemertines). (From the ecological point of view, on the other hand, this would not imply a divergence, since the usual division of the life cycle into a pelagic juvenile phase and a benthic adult phase would even in this case be the original condition.) The reason why I do not immediately exclude the possibility that such a life cycle could have been retained in certain polyclads is the fact that in the first multicellular forms of animals development must have been along the same lines. The turbellarians have, however, undergone a long period of evolution, and there is nothing to indicate that such primitive morphological features of the life cycle could have been retained. Since we know of several life cycles within the Polycladida where the Müller's larva has been lost, it appears most probable that this larva is the primitive form for the whole group.

With regard to the other groups of the turbellarians, it is still more difficult to arrive at a firm conclusion about details of the evolution of the life cycle, since they possess neither a Müller's larva nor any other special pelagic larval form. Here we find, direct development in all cases and, at least in the majority of the species, the life cycle contains no pelagic phase. In some forms and groups the life cycle became strongly modified in other respects (as a result of ectolecithal eggs, etc.). It is of course possible to reason that, in spite of the change to direct development, the juvenile form in some isolated cases has retained the original pelagic mode of life. Similar conditions are found, in the brachiopod genus *Lingula* (see p. 47).

It seems, in fact, that I myself have found an interesting example of this original mode of life. The following preliminary report is based on the study of living material.

During the months September to November there appears as a rarity, in the plankton of the western coast of Sweden, a kind of animal which must obviously be a juvenile form of some group of turbellarians. The animals are of an elongated oval shape and up to *ca*. 2 mm long. The largest specimen found was almost cigar-shaped. The smaller ones, up to a length of about 0·5 mm, were relatively thicker. The surface of the body is covered by fine and uniform ciliation. By means of it the larger animals after capture mostly creep about on the bottom of the dishes, while the smaller ones usually swim about quite swiftly. Some are colourless, others more or less intensely red or reddish violet. In this case the coloration is due mainly to small groups of minute grains of pigment in the epidermis. (The variations in the colour seem to indicate the existence of several different species.) The most noticeable feature about these young turbellarians is a statocyst situated at the anterior end and containing two spherical statoliths consisting of a hard substance. Pressure breaks them to sharp-edged fragments. No distinct mouth opening could be discovered; in a single case I believe to have found a trace of such an opening, roughly in the middle of the ventral side.

Statocysts with two statoliths are known in certain cases in the turbellarians, e.g. in *Nemertoderma* and *Meara* (Westblad, 1937, 1949). (I thank Prof. Tor G. Karling who has kindly called my attention to this fact.) In these two genera the statocyst is, however, constricted in the median plan. This could not be observed in any of my animals. The evidence nevertheless suggests that they are juvenile stages of nemertodermatids. It may be noted that several still undescribed species of *Nemertoderma* have recently been found upon littoral sandy bottoms (Sterrer, 1966).

Even if it is impossible to express any definite opinion about the organization of the primary larva of the last common ancestor of all groups of the platyhelminths, the most likely supposition is that in its essential features it was *trochophora*-like. This opinion is based on the embryogenesis and the preoral ring of cilia of the primary larva in the Polycladida, which from the ontogenetic point of view is the most primitive group. The last common ancestor might even have had a typical Müller's larva, but the possibility also exists that the characteristic lobes emerged only in the special evolutionary line leading to the Polycladida.

It has often been pointed out that in the Goette's larva the pair of lobes situated laterally from the mouth produces an appearance which is reminiscent of the *pilidium* larva of the nemertines, and for this reason it has been suggested that this similarity might be phylogenetically conditioned. Thus these lobes should be homologous to the "ear lobes" in *pilidium* (see e.g. Beklemischew, 1958, p. 126). This idea, however, presents considerable difficulties. Firstly it presupposes that the Goette's larva should be the original one for the platyhelminths and not the more common type with eight or ten lobes. This is very questionable and is not supported by any evidence. Secondly, it would mean that a typical *pilidium* is the original type of larva for all groups of the nemertines. This certainly is not the case (see p. 99). Finally, the two larval forms are so dissimilar in essential features of embryogenesis, morphology, and metamorphosis that the concept proposed is suspect on this consideration alone.

I might also point out that the "*lobophora actinotrocha*" (Fig. 19D) described by Dawydoff (1940a), in spite of its similarity with the larva of the *Phoronida* is quite unrelated to it in its phylogeny. The similarity is only superficial. It will be recalled that in the *actinotrocha* the tentacles are postoral, while the lobes in the Müller's larva in their original position are preorally situated (see Fig. 19A).

Finally, I must point out that the ontogeny of the polyclads—though primitive in some respects—has undergone considerable change. Thus no forms are known in which fertilization is external. Neither are the early embryonic stages free-swimming as is the case in, for example, certain nemertines.

Nemertini

With the exception of the holopelagic members of the Polystylifera which have secondarily changed their mode of life, the nemertines are benthic in their adult phase. The juvenile phase, on the other hand, is quite commonly pelagic in the three large subgroups.

Before embarking on a discussion of the different larval types I wish to recall that the ontogeny of the group contains some features which are original characteristics of all metazoans. These are: free discharge of the eggs into the water (not common; as a rule the eggs are held together by secretory envelopes of different kinds), external fertilization, ciliated and free-swimming coeloblastula, and gastrulation by invagination (cf. p. 239).

The morphology of the juvenile forms and their modes of life are very variable. And since the information in the literature is very incomplete on certain important points, it is not yet possible to form a complete picture of the phylogeny of the life cycle within the group. Sufficient is known, however, to permit certain important statements.

The Palaeonemertini, which are considered to be the most primitive subgroup, are generally said to have direct development. The pelagic phase is said to be relatively brief, and Thorson (1946, p. 149) supposes that during this period the young are lecithotrophic. Coe (1943, p. 209) seems to be of the same opinion. Because these suppositions do not agree with my own experience, I shall briefly present some of the more general results of my investigations.

The investigations were carried out at the Biological Station Klubban, Fiskebäckskil, of the University of Uppsala, material having been obtained by plankton netting in coastal waters. The results from Sweden were verified in all essential respects by investigations in the Bahamas (Bimini) and in southern India (Mandapam Camp), and may therefore be considered as representative for the group.

During late summer and autumn (August to October) fully-grown larvae of at least a dozen species belonging amongst others to the genera *Callinera*, *Tubulanus*, and *Cephalothrix* are quite common. The body shape varies from an elongated egg-shape to fusiform, or is in some cases almost cigar-shaped. The surface bears a dense and uniform ciliation. The mouth is ventrally situated, away from the anterior end (Fig. 20A). The species of *Cephalothrix* are characterized by a pair of eyes (Fig. 20B) situated in the epidermis. In some species they were found to be double. In at least *Callinera* an opaque belt of epidermal granules is found somewhat behind the middle of the body. Its anterior boundary is sharp, the posterior diffuse (Fig. 20A : 3). Within this belt a so-called lateral organ is found on either side, in *Tubulanus* and *Callinera* at least. The mesenteron cells contain an ample food reserve in the form of fat globules. The length of the body varies roughly from a little less than 1 mm to almost 2 mm.

The size of the body and the amount of stored fat gives the impression that the information about lecithotrophy may be incorrect. This impression became a certainty when a comparison between the young that had just left the egg and the fully-grown larva of *Cephalothrix* cf. *rufifrons* showed the ratio of the volumes to be 1 : 25, a very considerable amount of growth. Shortly afterwards I had the opportunity of witnessing the ingestion of food in two specimens belonging to two species, unfortunately still unknown.

These larvae proved to be voracious predators, feeding on other planktonic organisms of considerable size. On one occasion a larva was found to
devour a *pilidium* of a volume almost equal to its own. During this process

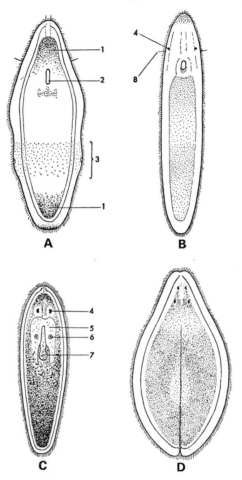

Fig. 20. Pelagic larvae of Palaeonemertini (A and B) and Hoplonemertini (C and D) from
the west coast of Sweden. A, *Callinera* sp., a fairly common larva, easily recognizable by
the intense coloration of the mesenteron: yellow, the foremost and hindmost parts being
red. B, *Cephalothrix* sp. Observe the position of the eyes in the epidermis. C, unknown
species with intense red pigment (dark in the drawing), mainly in the mesenteron. D, unknown species with feeble green pigmentation; dorsally and ventrally a marked furrow in
the median plan. Length: A and B, *ca.* 1 mm, C, *ca.* 0·6 mm, D, *ca.* 0·8 mm. 1, intense
red pigment in the mesenteron; 2, mouth; 3, belt of colourless granules in the epidermis;
4, eye; 5, brain; 6, statocyst; 7, proboscis; 8, sensory hairs.

the mouth was immensely widened. Before devouring the prey the larva rotated swiftly around it, its mouth apparently in contact with it. During this operation the prey may have been covered with an invisible secretion. Although the proboscis is already developed at this stage I could never observe that it was put to use. As far I can judge it is without function in the pelagic larvae.

My observations lead me to the conclusion that *the larvae of the palaeone-mertines are characterized by a long pelagic life and by planktotrophy*. The supporting evidence is, firstly, that the majority of the larvae examined become as large as, or even larger than those in which feeding could be directly observed, and secondly, that the alimentary system in all larvae I examined seemed to be organized in essentially the same way.

It is uncertain whether there are any exceptions to this rule. The scant information available in the literature supports my observations. Smith (1935) and Iwata (1960) report that larvae hatched in the laboratory and kept without a supply of food, after some time decreased in size, obviously from starvation. The greatest length attained by these larvae was never more than one third of the length at which my larvae, taken in the plankton, were transformed into the benthic form.

According to the observations by Smith (1935), Iwata (1960) and myself the newly-hatched larva has an almost spherical shape, and apically a strong tuft of sensory hairs. In the course of growth it becomes more or less elongate ellipsoidal, and the mouth, which in the newly-hatched larva is situated roughly in the middle of the ventral side, is shifted in a forward direction.

In his material Iwata could observe neither proboscis nor anus, whereas the proboscis was almost completely developed when my larvae were full-grown. (Unfortunately I have no information about the time of development of the anus.)

The transition to the adult phase is very simple. No parts of the body are shed, no radical transformation of organs takes place. All that can be observed after the descent of the larva to the bottom is that it becomes increasingly elongated and starts to creep about on the substratum in the same way as the adult, and that the catching of prey is now effected by means of the proboscis.

The phylogenetic interpretation of this transformation, which because of its uncomplicated nature hardly merits the term metamorphosis, is still uncertain. It is also doubtful whether the development of the palaeo-nemertines deserves the designation of direct development which is usually applied to it. It might be tempting to use this term if we take into consideration merely the unimportant morphological difference between juvenile and adult. The terminology is, however, less important than the fact that this subgroup in general still exhibits a pelago-benthic life cycle with planktotrophy in the pelagic phase. Future investigations must decide whether or not a more distinct type of larva was found in the ancestral forms (see also below).

Conditions in the larval phase which exhibit, at least superficially, a close agreement with the Palaeonemertini are also found within the Hoplonemertini (though not universally; see below). The shape of the body is simple, generally a more or less elongated oval (Fig. 20C), and the mode of life is usually pelagic. Some of the larvae which appear regularly on the west coast of Sweden differ at first sight from the larvae of the palaeonemertines only in the deeper position of the eyes (*underneath* the epidermis, not *in* it as is the case when eyes are found in the latter group, e.g. in the genus *Cephalothrix*; cf. B and C in Fig. 20). The transformation at the change-over to benthic life is again simple, entailing only a lengthening of the body.

In several cases, however, scientists have described more complicated conditions. Thus e.g. Dieck (1874), using " *Cephalothrix galatheae*"* and Delsman (1915) using *Emplectonema gracile* have observed a shedding of the larval ectoderm.† Information likewise exists pointing to the same process in certain other forms.

Here it may be added that the fresh water nemertine *Prostoma graecense*, which has eggs rich in yolk and direct development within the egg-envelope, and which consequently lacks a pelagic phase, according to Reinhardt (1941) exhibits features indicating a shedding of the larval epidermis once in the ancestry. This is evident from the fact that in the course of embryogenesis there appear solid agglomerations of cells which Reinhardt considers as equivalent to the blastodisks in *pilidium*.

Provided that the above statements and other similar ones in the literature are correct, it is logical to conclude that within the Hoplonemertini forms are found presenting conditions which resemble that type of life cycle within the Heteronemertini, where Desor's embryo occurs (see

* According to Coe (1943) this is more likely to be a species of *Carcinonemertes*. This is possible, *Cephalothrix* under no circumstances fits the observations.

† In studying Japanese material of what he considers as *Emplectonema gracile* Iwata (1960) was unable to verify this statement made by Delsman.

below). This in turn would indicate that a *pilidium*-like larva was found in the ancestral forms of the Hoplonemertini. In this case a change to direct development must have occurred in this group, in certain cases perhaps so completely that no traces of the *pilidium*-like larva are retained. (By using the term "*pilidium*-like" I do not mean that it necessarily has exactly the same body shape, but that it is a larva with embryogenesis and metamorphosis which agree in general with those in the *pilidium*; see also below.)

Further, in a Japanese species of the palaeonemertine genus *Tubulanus*, Iwata (1960) has found under the epidermis a number of fissure-like vacuities. He thinks that these might indicate that the epidermis is shed at the transition to benthic life. Even if it could be demonstrated that no such radical process takes place, it cannot be excluded that the fissures might indicate a shedding of the epidermis earlier in the evolutionary line leading to the Palaeonemertini, and consequently neither can it be excluded that a more complicated primary larva might have existed in the ancestral forms of all nemertines.

It should now be evident that the phylogeny of the larvae and larval development are somewhat obscure both in Palaeonemertini and Hoplonemertini. For this reason renewed detailed examinations of the different families and genera within these groups would be highly desirable.*

In spite of the fact that the Heteronemertini are usually included with the Palaeonemertini in the "class" Anopla their ontogeny is so far as is known entirely different, at least superficially. In Heteronemertini we encounter the remarkable *pilidium*-larva, remarkable especially for the well-known fact that after the pre-regeneration of the final epidermis the larval parts of the body are shed at metamorphosis.

In the three closely related genera *Lineus*, *Micrura*, and *Cerebratulus* the *pilidium* has been known for a long time. In the first two species there are, however, also three divergent types—Desor's larva (more correctly embryo), the "egg-swallowing type", and Iwata's larva.

Before discussing these divergent types I want to make the following remarks about the typical *pilidium*. This is a primary larva which through adaptation to pelagic life has acquired very individual morphology. For a

* One of my students (Cantell, 1969) has recently demonstrated the occurrence of *pilidium* in *Hubrechtella*. I cannot discuss here whether this should be taken to show that in the past the paleonenertines generally possessed a *pilidium* or that *Hubrechtella* does not belong to this group.

long time it lives as a free-swimming organism, and its food consists of particles much smaller than those ingested by the palaeonemertines (see p. 90). By the activity of the cilia of the surface of the body and of the stomodaeum the particles are carried into the mesenteron. Pronounced adaptation to the pelagic region has created a striking divergence from the adult. This in turn has brought about not only a far-reaching and fast metamorphosis, but has also necessitated certain other changes: a new epidermis and with it all the adult external organs and features are developed via the so-called blastodisks in the interior of the larva, so that they can assume their function immediately after the liberation of the young nemertine from the larval body and its transfer to the final biotope. The ingestion of food from outside by the liberated young nemertine may become somewhat delayed, partly because a certain amount of fat reserves are stored in the epithelium of the stomach during the larval period, and partly because, at least in some forms, the first meal after metamorphosis consists of devouring its own larval parts. I have observed this in several instances on the west coast of Sweden in young of a *pilidium gyrans* (see also Cantell, 1966a).

The features that should be considered most characteristic of the *pilidium*-larva are the developmentary ones described, rather than its external aspect—the helmet-like shape and the ear lobes. The reason for this reservation has been brought home by the description of new larval forms from Indo-Chinese waters by Dawydoff (1940a). He has shown that external appearance presents great deviations from what had long been considered characteristic of this nemertine larva. Since these discoveries by Dawydoff have not received sufficient attention in the literature, I shall give a selection from his pictures (Fig. 21).

In spite of the scarcity of information which is so far available about these different *pilidium*-larvae it should be safe to assume that all of them undergo the same fast metamorphosis as the "common" helmet-shaped larva. Thus there seems to be no essential variations in the development of the juvenile inside the larval body. In this respect the situation should also be the same in *pilidium recurvatum*, discussed below.

Here I should point out that some scientists use the term "metamorphosis" in a sense that differs from that adopted by me above. This is

---

FIG. 21. Different *pilidium*-larvae orientated in the same way: anterior lobe on the left. All should be planktotrophic. (After Dawydoff, 1940a.)

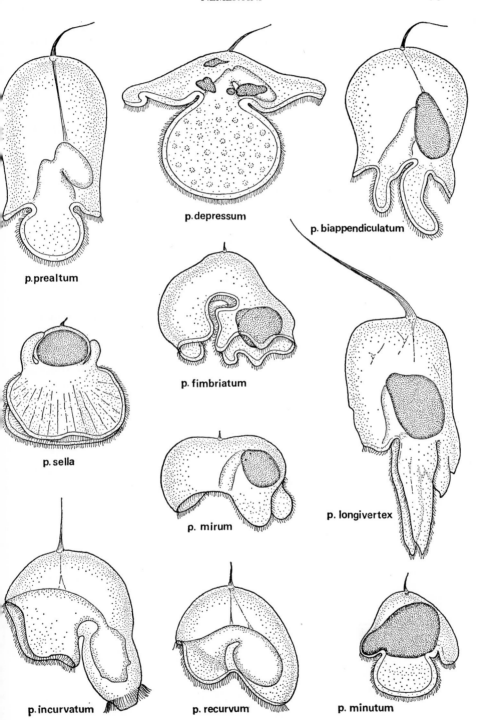

p. prealtum

p. depressum

p. biappendiculatum

p. sella

p. fimbriatum

p. longivertex

p. mirum

p. incurvatum

p. recurvum

p. minutum

evident from the following short remark by Iwata (1958, p. 130): "The young worm metamorphoses within the larval ectoderm." Here I should have said "develops" instead of "metamorphoses". In the nemertines as well as in the case of other groups with pelagic primary larvae I want to reserve the term "metamorphosis" for the (usually rapid) transformation which takes place at the change to the benthic mode of life, and which in *pilidium* essentially consists of the shedding of the larval parts.

A Desor's embryo occurs in the life cycle of a form of *Lineus* which appears in the literature under different names (*L. ruber*, *L. gesserensis*, *L. gesserensis-ruber*, etc.), but which is now called *L. desori* by Schmidt (1966). Here, in essentially the same way as in *pilidium*, the secondary (final) ectoderm is formed from the embryonic disks (blastodisks), while the primary ectoderm (the "larval skin") is lost. These developmentary processes should certainly be interpreted as a heritage from an earlier pelagic life. The hatching juvenile is essentially fully developed. The entire life cycle is benthic, and the part of it that corresponds to the *pilidium*-phase is compressed to a stage inside the egg. In other words, concurrently with the increased quantity of yolk the pelagic phase of the life cycle has been eliminated in the same way as in so many other cases within the metazoans. Thus, direct development has arisen, and it is therefore inappropriate to talk about a "larva" as is commonly found in the literature.

I want to stress that the free *pilidium* should be considered more primitive than Desor's embryo, as is now generally accepted. The development via Desor's embryo is direct development in all respects except that certain reminiscences after the *pilidium* are still very distinct. This probably explains why some authors have incorrectly called the development indirect. (If one were to speak about indirect development as soon as any traces of the primary larva are observable, the terminology would have to be changed in most cases.)

In a closely-related form which in his later papers he calls *Lineus ruber*, Schmidt (1934, and other papers) has proved the existence of a type of development which differs from that via Desor's embryo mainly by:

1. smaller size of the eggs,
2. the fact that only about one-third of them develop in the ordinary way, and
3. the fact that those eggs which do not continue to develop (abortive eggs) are eaten by the normal embryos.

Schmidt is of the opinion that this "egg swallowing type" is derived from the Desor-type.

On the basis of our present knowledge I am not prepared to exclude the possibility that the Desor-type has given rise to the egg-swallowing type, but I cannot share Schmidt's opinion that this latter has arisen "infolge einer Verminderung der Eigrösse und Erhöhung der Eizahl", and that the swallowing of eggs has appeared only *after* this decrease in egg size. It is also very difficult to imagine that the egg-swallowing could have emerged in young more or less undifferentiated embryos (in order to compensate the reduced yolk). I find it more possible to visualize evolution taking place in such a way that ingestion of the eggs began in fully-developed young (in an ontogeny with a typical Desor's embryo), but was then successively shifted to ever younger embryonic stages.

Another, but perhaps less probable explanation could be based upon the assumption that Schmidt's "larva" is derived directly from a *pilidium* enclosed in an envelope, and that this *pilidium* continued to take up food after its "imprisonment" in the same way as during its earlier pelagic life, but with the change that now abortive eggs instead of plankton were eaten. I do not think it feasible to exclude this possibility out of hand since we know of such a case of evolution among the gastropods (see p. 127 *et seq.*).

Iwata (1958) has described yet another variant within the Heteronemertini—a pelagic larva in a Japanese species of *Micrura* (*M. akkeshiensis* Yamaoka). This so-called Iwata's larva (Fig. 22) differs from a typical *pilidium* mainly by:

1. absence of all lobes so that an external radial symmetry is produced,
2. orientation of the enclosed young in such a way that its anterior end is situated at the posterior end of the larva, and
3. lecithotrophy.

In other respects it agrees with the *pilidium* in its shedding of the larval skin after formation of the adult epidermis in the usual way via the blastodisks (amniotic invaginations). (No uniform amniotic envelope enclosing the whole embryo as in the *pilidium*, however, is formed. In Iwata's larva this envelope has obviously been reduced.)

Iwata says that "the present larva corresponds to a Desor larva adapting itself to swim during development. If we push a step further, we may obtain a pilidium larva which swims and takes in food" (1958, p. 129). This remark makes it evident that the Japanese scientist is of the opinion that evolution has taken the following course: *Desor's embryo → Iwata's larva → pilidium*. To me such a sequence of evolution appears entirely impossible (cf. what has been said above about Desor's embryo). A nearer approach to the truth is to suppose that the course followed was the opposite. The evolution from *pilidium* to Desor's embryo certainly contained an intermediate lecithotrophic stage. It was probably a free-swimming larva which perhaps resembled Iwata's larva, but it is hardly

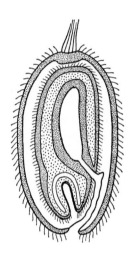

FIG. 22. Diagram of Iwata's larva. (After Iwata, 1958—simplified.)

possible that Iwata's larva, which is known only in one of the species of *Micrura* (in the others only the *pilidium* is known), is the connecting link. There is, however, no doubt in my mind that both Desor's embryo and Iwata's larva have arisen from the *pilidium*-type. Of the two life cycles, that with the Iwata's larva is the less modified, *inter alia* in the respect that it has retained the pelagic phase.

A very interesting larva in the Heteronemertini is *pilidium recurvatum*, discovered by Fewkes (1883) at Newport on the east coast of North America. Since then, as far as I know, no further find of this larva has been reported in the literature, but one of a very closely related form (*p. incurvatum*) by Dawydoff (1940a). However at the end of August 1962 I found a specimen on the western coast of Sweden (Gullmar fiord, in surface water), which I was able to study in a living condition, and this proved to agree with *p. recurvatum* in all essential points. The investigations have been continued by Cantell (1966b) who has found a few further specimens.

Pilidium recurvatum differs considerably from a typical *pilidium* (in the restricted sense) by the general shape of its body and by the ring of cilia at the posterior end (Fig. 23A). Despite these differences, however, there occur essential similarities, especially as regards the development of the embryo inside the larval body. The occurrence of an amnion makes it apparent that the final epidermis is formed in the same way as in a typical *pilidium*, and that the special larval parts of the body are lost at metamorphosis. Fewkes' description of the process is, however, not quite clear.

Unlike Fewkes I do not find it difficult to compare *recurvatum* with a typical *pilidium*. This is mainly because I have found a posterior slit in the bent part (the "funnel"). This part is thus not a completely closed tube as suggested by Fewkes' illustrations. It might also be pointed out that the larval mouth is not situated terminally on the funnel but rather subterminally and between two lateral lobes of an almost triangular shape. The position of these lobes in relation to the mouth is thus exactly the same as that of the "ear lobes" in an ordinary *pilidium*, and they are doubtless homologous to these structures.

After giving these explanations the rest is self-evident. The major part of the body with the ring of cilia corresponds to the posterior lobe of a typical *pilidium*, while the anterior marginal portion of the funnel corresponds to the anterior lobe. I refer to the series of diagrams A–C in Fig. 23 which explain the relationship better than words. From these diagrams it appears that apart from the ring of cilia in *p. recurvatum* it is only a question of differences in proportions. These differences account for the fact that the orientation of the embryo enclosed within the larval body differs by about 90° in the two types.

It is interesting to note that Dawydoff (1940a) who follows a different line of thought arrives at essentially the same conclusions as to the homologization of the parts of the body in his *p. incurvatum*. In the comparison he was aided by the find of *p. recurvum* which he himself believed to be a young *p. incurvatum*. This find led Dawydoff to the conjecture that evolution had progressed from the ordinary *pilidium* type to the *recurvatum* type.

I find this interpretation of the evolution by Dawydoff improbable, and consider the reverse to be more likely. It is obvious that the ordinary *pilidium* is a highly altered larva.

The *recurvatum* type might be said to be more "normal". In its external appearance it differs from a larva of Palaeonemertini mainly by its ring of cilia and by the funnel which in fact is nothing more than strongly extended and enlarged perioral parts.

Even if I dare not derive *p. recurvatum* from the type found in the Palaeonemertini (cf. p. 89), it is, at all events, certain that it has arisen

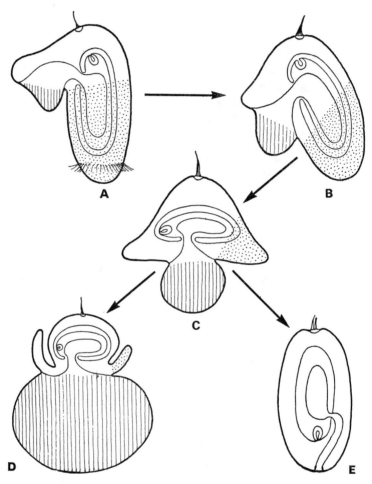

FIG. 23. Series of diagrams to show the probable evolutionary connections between some of the more marked types of pelagic larvae within the Heteronemertini. A, *p. recurvatum*; B, transitional form, almost agreeing with *p. recurvum*; C, typical *pilidium*; D, *p. sella*; E, Iwata's larva. In order to facilitate comparison the "ear lobes" have been marked with vertical lines, the posterior lobe with dots. Note the different orientation of the embryo within the body of the larva.

from a simpler type which had no funnel and thus possessed a "normally" situated mouth, and which did not shed any larval parts of the body at metamorphosis.

If the proposed interpretation is correct, a diagram illustrating the evolution of some of the more important larvae would have the appearance as shown in Fig. 23. *Pilidium sella*, especially, is radically altered; it has enormous "ear lobes" (according to my interpretation above these are only the extended lateral portions of the mouth, and thus really lateral lips), and strongly reduced unpaired lobes which are bent up towards the apical pole. Strongly altered also is Iwata's larva, which has lost all lobes and where in the course of the evolution from the *recurvatum* condition the embryo has turned through 180° inside the larva.

It deserves to be mentioned that the two unpaired lobes which in many pilidia are almost, if not completely similar, must have evolved from extremely unlike and non-equivalent parts of the body of the original type: the anterior lobe is only an enlarged and somewhat displaced preoral part of the mouth, a type of "upper lip", while the posterior is identical with the entire postoral body of the larva. Because these parts, which are genetically so dissimilar, have in many *pilidium*-forms assumed the same size and shape, the larval body has become markedly bisymmetric (biradial). It is remarkable that in certain *pilidia* the bisymmetry is expressed even in such details as the number and localization of the frequent agglomerations of small pigment grains in the margins of the lobes. The acquisition by Iwata's larva of an external radial symmetry provides further evidence for the course of evolution: bilateral symmetry → bisymmetry → radial symmetry found here and there within the metazoans (cf. *Priapulus*, p. 74).

It should be stressed that, quite irrespective of what has just been said about the direction of the evolution, *pilidium recurvatum* cannot have any closer phylogenetic connection with *actinotrocha*, *tornaria*, or with the larvae of echinoderms as proposed by Fewkes. The similarities adduced by Fewkes (1883), on which I do not think it necessary to comment further, are nothing more than features of superficial convergence.

Referring back to Palaeonemertini and Hoplonemertini I must point out that much still remains to be done before we can discern the phylogenetic traits behind the larvae and their ontogeny in all groups of nemertines. It has nevertheless been possible to throw light upon some important questions, and I hope that the discussion may in part have contributed to the formulation of the problems.

In conclusion I want to underline that the original division of the life cycle into a pelagic juvenile and a benthic adult phase has been retained in all three groups of nemertines in spite of the great variations in their ontogeny. Also, in the nemertines (inasmuch as they have or have had a *pilidium*-larva), radical transformation of the primary larva in connection with its adaptation to pelagic life has produced a strong divergence between the two phases of the life cycle. This divergence has eventually resulted in a far-reaching metamorphosis involving first of all the shedding of great parts of the larval body. Evolution has, however, in certain cases resulted in the elimination of the pelagic larva, and its replacement by benthic direct development.

It is finally worth mentioning that in the larvae of the nemertines no ancient adult characters seem to occur. Besides, as far as the *pilidium* type is concerned, no recent adult characters are displayed in the external morphology. As all external parts are lost at metamorphosis, the occurrence of such characters in the larva would be altogether meaningless.

Entoprocta

All entroprocts possess free-swimming larvae. Furthermore, the group is, without exception, characterized by a sessile mode of life in the adult stage. Of the three families distinguished the Pedicellinidae and Urnatellidae are permanently attached to the substratum and are colonial. The species within the Loxosomatidae are admittedly capable of limited locomotion, but by no means free-living in the usual sense of the term. Also their entire organization bears the stamp of sessility.

All species, or at least all that have been examined, are known to possess inside the vestibulum a special place for brooding, where the embryos are attached for some time by means of a secretion. While in the brood chamber the embryos increase considerably in size, and for this reason it is apparent that they are provided with food in some way. For *Barentsia* Mariscal (1965) reports that the bigger larvae actively take up food from the atrial groove of the mother animal.

According to current belief the larvae, after leaving the brood chamber, swim about freely in the water. This free life lasts only a short time during which the larvae do not take in any food. There can be no doubt that this is the case in certain species, and it may even be general for species in temperate waters. However, it has recently been possible to establish a long pelagic life and planktotrophy in larvae belonging to two somewhat different types (Figs. 24 and 25). These observations have been

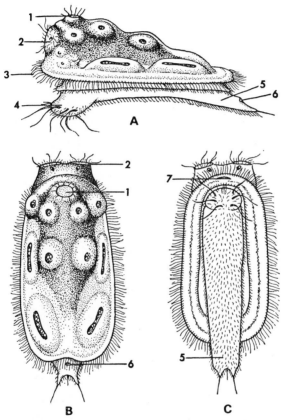

FIG. 24. Larva of entroproct, *Loxosomatidae*, with well-developed creeping organ (foot), drawn in creeping attitude. A, seen from the left; B, dorsal view; C, ventral view. On dorsal side, five pairs of glandular organs with round or elongated openings. 1, apical organ; 2, frontal organ; 3, prototroch; 4, foremost part of the foot; 5, hindmost part of the foot; 6, anus; 7, mouth. (After Jägersten, 1964.)

made on material from the Bahamas (Jägersten, 1964). And since I have also found similar larvae in southern India (Mandapam Camp), it appears as if such conditions may be more common in the warmer regions. Larvae related to that shown in Fig. 25 have also been described by Nielsen (1966) and Franzén (1967).

In the literature the entoproct larva has usually been interpreted as an admittedly slightly modified *trochophora*. A divergent opinion is held by

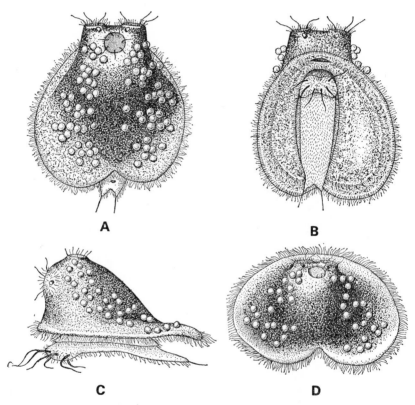

FIG. 25. Entoproct larva of a type with internal budding. A, dorsal view; B, ventral view; C, seen from the left, creeping; D, in dorsal view, swimming. (From Jägersten, 1964.)

Hyman (1951b) who finds the larva of *Pedicellina* to "bear no great resemblance to a trochophore either superficially or as to structural details", and who asserts that that of *Loxosoma* is more reminiscent of a rotifer.

On the basis of knowledge acquired by my own investigations I must reject Hyman's concept. The similarity with a rotifer is slight and finds its expression only in the fact that the creeping organ (see below) is divided posteriorly into two points or lobes presenting a superficial similarity with the so-called "toes" of the rotifers. It must, however, be pointed out that these lobes are entirely devoid of adhesive glands or mechanism for fixation. Neither do they agree with the toes of the rotifers in any other detail.

Neither is there any closer agreement with the rotifers in early embryogenesis, but there is good agreement with annelids and molluscs. We find pronounced spiral cleavage, coeloblastula, gastrulation by invagination, formation of the mesoderm via mesentoblast cells, and mesoderm bands.

After the true metatroch is established (Jägersten, 1964) a close agreement between the ciliary covering of the entoproct larva and that of the original *trochophora* type is obvious. It is of particular interest that this similarity holds also for that part of the ciliary apparatus which is involved in feeding, i.e. the adoral ciliated zone (see below, p. 159).

As in the larvae of several other groups (see p. 244), a longitudinal band of cilia is found between mouth and anus, and it is by means of this band that creeping locomotion on the substratum takes place (cf., e.g. the larva of *Protodrilus*, Jägersten, 1952). It is of particular interest that in the entoproct larva the portion of the body that carries this band of cilia is differentiated as a special creeping organ—a "foot" that greatly resembles the foot of a gastropod (Figs. 24 and 25). The possible interpretation of these agreements will be discussed below (p. 247).

There is another reason why the occurrence in these pelagic larvae of a well-differentiated creeping organ is of great interest. It is obvious that the foot is not a larval organ in the sense that it fills any function during the pelagic phase of development. At least in the type which gives birth to young and then dies (see below) it is never put to use any in the natural environment. This means that there is no justification for the assumption that the foot represents an adaptation to the pelagic mode of life.

Since, furthermore, the foot is missing in the adult (which in its present transformed condition has no use for such an organ of locomotion), it can only be interpreted as an ancient adult character. This indicates that the adult of some ancestral form was able to move about freely as a creeping bottom-dwelling animal (Fig. 28A).

(The creeping foot in the entoprocts is paralleled in such parasitic gastropods as *Entoconcha* and others. In the adult stage these forms lack any trace of the characteristic foot in common gastropods, but their larva exhibits a more or less distinct remnant of this organ. Thus in these parasites the foot without doubt is an ancient adult character.)

In some creeping ancestor, "adult pressure" (see p. 6) had caused the development of the foot to shift to the larva, perhaps even to a stage in embryogenesis, in this way simplifying metamorphosis. After the transition of the adult to sessile life the foot was reduced in the adult stage, finally disappearing, but was retained in the larva, where it probably still functions in moving to a suitable place for settling in those forms that still metamorphose into the adult in the original way.

It can be taken for granted that the shifting of the development of the foot to the larva took place at the time when the adult was still free-living. No such transition was possible once the adoption by the adult of a sessile mode of life had led to disappearance of the foot in the adult stage, since the necessary adult pressure was now lacking (cf. p. 222).

Here it may be mentioned that several specimens of a larva entirely without foot have been encountered (Jägersten, 1964, and later unpublished finds). Unfortunately we still know nothing about the earlier stages of such specimens, but whether the foot is never developed or is formed only to be eliminated early on, it is still probable that this lack is an indication of a general reduction of the foot in the larvae of the entoprocts. This is also what might be expected. Just as once the development of the foot was shifted to the larva as the result of adult pressure in the past, the foot is now in the process of being eliminated by contemporary adult pressure, since it is no longer found in the adult. The same phenomenon can be observed in other forms, in connection with other negative characters (see e.g. pp. 59 and 123).

Vegetative propagation by budding is a recent adult character that has been shifted to the larva. Such propagation occurs in the adults of all known entoprocts, and there should be no doubt that, as in the case of other sedentary groups with this type of propagation, it has only arisen after the appearance of the sessile mode of life. Vegetative propagation in the larva (see Jägersten, 1964) cannot be considered as an entirely new phenomenon which has nothing to do with budding in the adult, and we must consider larval budding as a result of adult pressure. Thus the ability of forming buds appeared, as in the Bryozoa (see p. 45), first in the adult, but later on also in the larva. We still do not know how widespread larval budding is within the group, but to judge from the detailed investigations carried out by outstanding scientists on *Pedicellina*, it does not occur in this genus.

The shifting of the process of budding to the larva was followed by certain changes. In one type of larva at least (type B; Jägersten, 1964), the budding has been displaced to the interior of the body (Fig. 26C). This can probably be considered as an adaptation to the pelagic mode of life, since large buds projecting from the surface of the body would be a considerable obstacle while swimming.

Somewhat later in the evolutionary line leading to these forms with internal budding another change appeared, so that the larvae no longer

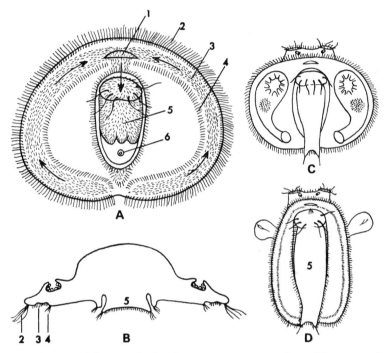

FIG. 26. A, diagram of the ventral side of a swimming entoproct larva. The arrows show the transport of food particles to the mouth. B, outline of transverse section through swimming larva. C, larva with two enclosed young, ready for birth. D, larva of another species with two external buds. 1, mouth; 2, prototroch; 3, adoral zone of cilia (atrial groove); 4, metatroch; 5, creeping organ (foot); 6, anus. (After Jägersten, 1964.)

underwent metamorphosis, but died off after having released the offspring (see Jägersten, 1964, p. 311). This might alternatively be expressed in the following way: since the transformation of the larva into the adult was now unnecessary, this original process of development could be cut out without any disadvantageous results for the continuance of the species. It might be supposed, however, that this change only took place after a period of evolution during which the larva was able both to develop buds and to metamorphose in the ordinary way.

Connected with these changes is the fact that larvae with internal budding, contrary to the general rule, no longer store any food reserves in their stomach cells. Such reserves in larvae which will in any case die, would be quite unwarranted.

Another feature in the larva which one might be inclined to interpret as a recent adult character is the "atrial groove" (Fig. 26A). This has the same morphology and function as in the adult stage (cf. Atkins, 1932 and Jägersten, 1964). However, I do not consider that the groove is an adult character the development of which has been shifted ("accelerated") to the larva, but rather a larval feature which has been retained in the adult after its alteration by neoteny. I hold this opinion because the feeding mechanism in the entoproct larva is identical with that in those larvae of annelids and molluscs which are primitive in this respect (see pp. 160 and 121; cf. also the discussion of the rotifers on p. 67).

All the indications are that this feeding mechanism is very ancient, having already arisen in the larva of the ancestral form common to Entoprocta, Annelida, and Mollusca, perhaps also to Rotatoria and other groups. The mechanism is an adaptation to pelagic life, but it also fits the benthic adult of the entoprocts, and could therefore be retained in this case. The similarity here is thus not the result of adult pressure.

The location of the brain in the entoproct larva may on the other hand be a result of adult pressure, in this case an ancient adult pressure. In contrast to many other larvae the brain is not epidermally situated, but is entirely inside the connective tissue. In spite of its transitory nature (it is lacking in the adult), the brain exhibits a well-developed histological differentiation, as do all tissues of the larva with the exception of the buds and the so-called x-structures (the latter can probably be classified as buds too; see Jägersten, 1964, footnote on p. 312, and Nielsen, 1966, p. 227).

New facts established about the larvae of the entoprocts and the comparison between them and their adults by applying the principle of adultation have thus thrown more light on past and present conditions in the group. Some additional remarks remain to be made about ancient features.

We are forced to assume the existence of a ciliated creeping foot in the adult form of the free-living ancestors. This was in all probability much the same as the organ in the larvae of Loxosomatidae (Jägersten, 1964). No detail can be provided about the organization of this adult in other respects. It seems, however, that we should distinguish between two different adults during the period of non-sedentary life, since it seems that neoteny had taken place already at an early evolutionary stage (see below).

No essential differences can have existed between the larvae of the free-living ancestral forms and the recent larva. The initial development of the creeping foot had been incorporated into embryogenesis during the free-living evolutionary period, and because of this the metamorphosis of the larva into the benthic adult was relatively simple at that time.

The divergence between larva and adult became considerably greater once the latter had adopted a sessile mode of life and consequently had been altered in many respects, *inter alia* had lost the foot. This resulted in more pronounced metamorphosis, but at first not as far-reaching as the metamorphosis which now takes place in *Pedicellina* and which we know mainly from Harmer's (1886) and Cori's (1936) investigations.

In *Pedicellina* metamorphosis follows roughly the following course (Fig. 27): the larva fixes itself to the substratum with the margin of the body just outside the infolded prototroch, and the margin then grows in-wards and closes with the result that the entire ventral area with foot, mouth, and anus is entirely shut off from the surroundings. Then the part of attachment becomes elongated into a compact stalk, and simul-taneously with the reduction of the purely larval organs all the internal structures rotate by about half a turn, After this the enclosed space (atrium) opens, and the tentacles, which in the meantime have started their development inside the atrium, as have the mouth and anus, come into contact with the outside world.

It is quite obvious that this metamorphosis does not recapitulate the phylogenetic changes in the adult. In no generation have the crown of tentacles, mouth, and anus been without contact with the water for the simple reason that they could not have functioned. For this reason we must imagine that the first sessile ancestral forms were attached only by a pre-oral part of the anterior end, and that the stalk has evolved from this part.* We should also consider that at first the sessile condition was only temporary, alternating with longer or shorter periods of creeping loco-motion.

* This concept receives direct support from the fact that in a larva with external buds (Fig. 26D) these were situated antero-laterally, thus in the place which most closely corresponds with the insertion of the stalk in the calyx of the adult, and also where the adult develops its own buds (cf. footnote on p. 114).

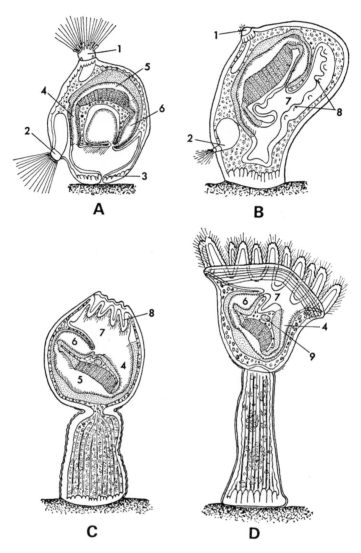

Fig. 27. *Pedicellina cernua*. Four stages of metamorphosis. A, larva just after settling; atrium about to be closed. B, stalk starts to form. C, stage after development of stalk. D, after metamorphosis; the atrium has opened and the tentacles have been bent outwards. 1, apical organ; 2, frontal organ; 3, margin of body (the prototroch is already lost); 4, oesophagus; 5, mesenteron (stomach); 6, rectum; 7, atrium (vestibulum); 8, developing tentacles; 9, abdominal ganglion. (After Cori, 1936.)

Fig. 28 shows the evolution of the adult as I see it. As regards the initial stage, the free-living ancestral form (A), the only thing that can be said with certainty is that it possessed a creeping foot which on the whole resembled that of the recent larva, and that it was devoid of tentacles. It is, however, probable that in common with the recent larva it also

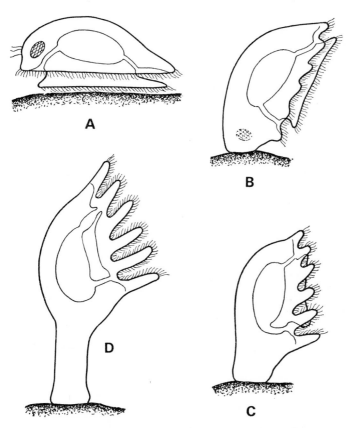

FIG. 28. Diagrams representing the probable course of evolution in the *Entoprocta*. A, free-living ancestral form; B, stage soon after adoption of sessile mode of life; C, further changed stage; D, recent adult (very close to *Loxosoma*). See further on the text.

possessed the ciliary feeding apparatus. For this reason I have drawn a prototroch at the margin of the body in Fig. 28A, and we should imagine that metatroch and adoral cilia, too, were present (cf. Fig. 26A). In the next stage (B), which had fixed itself with the preoral portion in front of the prototroch, I think the first indication of the tentacles might have been found, and that in the following stage (C) these structures were fairly well developed. The series of diagrams B–D also shows not only the elongation of the anterior end into a stalk but also the reduction of the creeping foot.

It is not without some hesitation that I conceive of a replacement of the unknown feeding apparatus in the adult free-living ancestor by the characteristic larval ciliary apparatus at this early stage of evolution. It cannot be excluded that this change occurred only simultaneously with the adoption of a sessile mode of life. The sessility did not appear abruptly, however, and it is therefore more probable that the larval ciliary apparatus was taken over by the adult when this was still free-living. At all events, the larval feeding apparatus was a feature in the adult by the time the sessile mode of life had been fully adopted.

The following comments may be made about the phylogenetic appearance of the tentacles. In a paper in 1964 I established the close agreement between larva and adult over the feeding apparatus, if one disregards the fact that the tentacles have taken over the function of the prototroch. I now feel justified in carrying this comparison a step further.

The background is supplied by our present knowledge that the ciliation of the tentacles is such as to permit comparison with the prototroch, as well as with outer portions of the adoral ciliation. The lateral cilia of the tentacles function in the same way as those of the prototroch for catching food particles, and the "frontal cilia" (see Fig. 29B and C) then transport the particles towards the "atrial groove". It appears, therefore, that the tentacles should be considered as processes from the margin of the body, essentially from the region of the prototroch. In their formation the prototroch has been extended into an arc-shape at every tentacle, and a portion of the adoral ciliation has become situated inside this arc (Fig. 29C).

This interpretation implies that in the entoprocts the ring of tentacles is *preoral*, as has already been pointed out (e.g. Cori, 1936). The point is worth stressing, since the crown of tentacles of the bryozoans and the other tentaculates is *postoral*. It might be added that the water-current produced by the tentacle cilia in the entoprocts (Fig. 29A) is in the opposite direction to that in the tentaculates. This is explained by the circumstance that the former are attached to the substratum by a preoral part of the body, while the latter are fixed by a postoral (ventral) region (see p. 52).

The loxosomatids have not moved very far from the conditions which

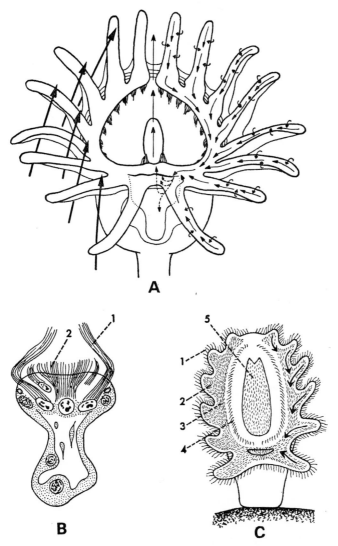

FIG. 29. A, crown of tentacles in *Loxosoma*. Large arrows to the left indicate the direction of the current of water between the tentacles. The smaller arrows to the right show the transport of food particles to the mouth. (After Atkins, 1932—modified.) B, cross-section of tentacle. (After Atkins.) C, imaginary stage in evolution of the adult (cf. C in Fig. 28), seen from the ventral side and exhibiting the ciliary apparatus for catching and transporting food particles (cf. A and B). 1, prototroch (=lateral cilia of the tentacles); 2, "frontal cilia" (=adoral cilia of the tentacles); 3, adoral ciliated zone (=atrial groove); 4, metatroch; 5, remnant of the creeping foot.

existed immediately after the transition to the sessile mode of life. The stalk is relatively short and may even be missing in certain forms. (In this case it might, however, represent a return to more primitive conditions.)

The stalk still issues from near the anterior end, and the animal is still able to detach itself occasionally from the substratum. The original type of locomotion has, however, disappeared.

In the Pedicellinidae and the Urnatellidae the change has progressed further. Here the stalk issues from the area which corresponds with the dorsal side of the larva, and vegetative propagation has resulted in the development of permanent colonies.

Although I arrived at the conclusion that the ancestral form attached itself by its anterior end, this does not compel me to believe that its larva necessarily did the same at metamorphosis. At first the larva may have fixed itself by the whole margin of the body in the same way as the recent larva in *Pedicellina*. This is possible provided that within a short time the individual shifted the point of attachment to the anterior end. (This would be necessary to facilitate feeding.) There is, however, no reason to suggest that the larva of the first sessile ancestor did not fix itself directly by the anterior end. For this reason the evidence suggests that the present course of metamorphosis is more complicated than the ancient one. In making this statement I refer of course to the only metamorphosis which is known, viz. that in *Pedicellina*. It is possible that in other forms simpler conditions exist, and for this reason renewed investigations would be most desirable.*

I am not the first to arrive at the opinion that the ancestral form of the entoprocts fixed itself by the anterior end when it became sessile. This view has been held by e.g. MacBride (1914) and by Brien and Papyn (1954).

It is now possible to conclude that the brain of the ancestral form, like the preoral organ, provided such a structure also existed in the adult, has been lost. (Thus the disappearance of the larval brain at metamorphosis recapitulates a phylogenetic process.) Consequently the ganglion which in the adult is situated below the bottom of the atrium is not a brain but an abdominal (subenteric) ganglion, as was first maintained by Hatschek.

* Since the above was written, Nielsen (1967) has, interestingly enough, found that in one species of *Loxosomella* the larva, which becomes transformed in its entirety into the adult, fixes itself by the frontal organ, and that the stalk of the adult is developed from the preoral portion. This can be considered to prove that the opinion expressed by me here is the correct one.

In my conclusions I am joining those scientists (e.g. Cori, 1936; Brien and Papyn, 1954), who deny that there is a close relationship between Entoprocta and Bryozoa. The similarities between these groups, in most respects very superficial, can all be attributed to convergence as a result of the sessile mode of life. The established similarity between the larvae of entoprocts, annelids, and molluscs with regard to the ciliary apparatus for feeding has made this still more obvious. I must thus emphatically reject Marcus' (1938, 1939) attempt to vindicate the older concept. Marcus' opinion is based on an erroneous interpretation of the polarity in the larva of the *Phylactolaemata* (see p. 35 *et seq.*). Compare also what has been said (p. 112) about the position of the tentacles and the direction of the water current created by them.

In the above I have arrived at the conclusion that the entoprocts are neotenic in so far as the adult has retained the larval ciliary apparatus for feeding. There is, however, the possibility that in the evolutionary line of the entoprocts, while the adult was still free-living, total neoteny took place, i.e. that the existent larva became sexually mature and that the corresponding adult was eliminated. In such a case it would have been the new (secondary) adult that later gave rise to the recent, sessile one.

Such an interpretation is supported by the close agreement between larva and adult at the present time, which applies to almost all points except the tentacles and the stalk of the adult and the creeping foot of the larva. This agreement between the two developmental stages has been pointed out already, *inter alia* by Mariscal (1965) who appropriately designates the fully-developed larva as a swimming calyx.

From the idea that the recent entoprocts might be the descendants of sexually mature larvae, it follows that we must reckon with the *possibility* that the original, free-living adult, i.e. the adult which was eliminated when the larva became sexually mature, might have been a highly organized animal. There is nothing to prevent it from having possessed a spacious coelom, metanephridia, and perhaps even a circulatory system, that is to say systems of inner organs which as a rule are developed at or after metamorphosis. The only distinct organ retained in the recent larva reminiscent of the eliminated adult is the creeping foot discussed above. The retention of the foot is due to the fact that its initial development was transferred to the larva. This was not the case with the other organs just mentioned.

By way of conclusion I shall summarize the most important of the discussed changes in evolution terminating in the recent entoprocts:

1. In the benthic phase of the life cycle of an early ancestor, which moved on the substratum by means of a longitudinal ventral band of cilia (cf. p. 244), a distinct creeping organ (a "foot") was differentiated. The planktotrophic larva was of the original *trochophora* type.
2. The initial development of the foot became transferred to the pelagic larva.
3. The adult was eliminated by neoteny. The new larva-like (secondary) adult was, or soon became benthic.
4. The new adult attached itself by the preoral part which became elongated into a stalk. The sessile mode of life led to the loss of the foot, and also the rest of the body became considerably altered. Among the changes were the evolution of tentacles and vegetative propagation by budding.
5. In some forms the budding was extended to the pelagic larva.
6. In some forms this larval budding became internal.
7. In these forms the development was changed in such a way that the larva no longer metamorphoses into the adult but dies after having "given birth" to the individuals formed by budding. (This situation still occurs at the present time.)
8. In certain recent larvae the foot has disappeared.

Mollusca

In the molluscs both holopelagic and holobenthic life cycles occur as well as the original one with pelagic primary larva and benthic adult. The last-named cycle is strongly dominant; it occurs in all groups of molluscs, known for a long time, with the exception of the Cephalopoda. The larval development of Monoplacophora is still unknown.

For the sake of clarity, I want to point out right away that the ontogeny of the cephalopods differs radically from that of the other groups. They have large eggs, very rich in yolk and, correspondingly, peculiar embryogenesis and direct development inside the envelope of the egg. Conditions are so strongly altered that no trace of the primary larva is found. For this reason the cephalopods can be completely disregarded in what follows.

They do, however, make us understand how radically and relatively quickly the ontogeny of a group of animals can be changed.

Within Polyplacophora, Solenogastres, Gastropoda, Lamellibranchiata, and Scaphopoda the original type of ontogeny is more or less frequently represented. Thus in all five groups cases are found where the sexual products are discharged freely into the water, where the sperm is of a type that is original for the metazoans, and where, correspondingly, fertilization is external (Franzén, 1955). In all these groups the embryogenesis includes coeloblastula and invagination gastrula, and in certain cases

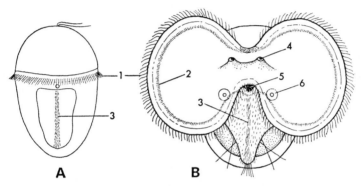

Fig. 30. A, diagram of the *trochophora* of *Ischnochiton* (Polyplacophora). (After Heath, 1899—simplified.) B, *veliger* of a nudibranchiate. (In part after Thompson, 1959.) 1, prototroch; 2, metatroch (seen through the velar lobe); 3, band of larger cilia on ventral surface of the foot; 4, eye; 5, mouth; 6, statocyst.

(this applies at least to a number of lamellibranchs) even the blastula is covered with cilia and swims about freely—a very primitive feature. (About original features in the ontogeny, see p. 239 *et seq.*)

Often the primary larva is still an ordinary *trochophora*, and it has long been believed that this larva was found in the ancestral form of all molluscs, In my opinion this view must be adhered to, mainly because of the often close agreement of the embryogenesis and morphology of the larva with other *trochophora* groups, particularly the annelids. (I must reject Riedl's doubts as to the correctness of this view, see p. 143.)

A virtually unaltered pelagic *trochophora* is, however, not the rule for the molluscs. An externally fairly typical *trochophora* does occur in a number of cases, but then it is either lecithotrophic and has only a short free-swimming life—as in Polyplacophora (Fig. 30A) and certain Solenogastres—or it has been altered into a *veliger*-larva.

In the majority of cases the *veliger*-larva is already fully developed when it hatches from the egg. Thus the *trochophora* is reduced to a stage in embryogenesis, as has often been the case with e.g. *nauplius* in the *Crustacea*. In many cases the *veliger*-larva has in turn become an embryonic stage with the result that the pelagic phase of the life cycle has been entirely lost (see below).

I want to stress that in fact the *veliger*-larva is nothing more than an older *trochophora* in which the ciliated oral zone has been extended so as to form the velar lobes. Thus the *veliger*-larva must not be considered a secondary larva (see definition, p. 8).

The oral ciliary rings as they occur in the primitive *trochophora*, as well as the entire velar lobes in the *veliger*-larva, are purely larval features, i.e. adaptations to a pelagic mode of life. As such they are lost at metamorphosis (see below).

In the molluscs, as in so many other groups of animals with primary larvae a shifting of the initial development of certain adult characters to the *veliger* stage has taken place, in some cases even to the *trochophora* stage or still earlier stages. The mantle, shell, and foot, and in the gastropods often the operculum, are the most marked external characters that have undergone this "acceleration".

The acquisition of these adult characters causes the larva to assume such a similarity with the adult—this applies especially to the prosobranchiate gastropods and the majority of the bivalves—that it would almost be feasible to speak of direct development, provided the velum is not taken into account. In other words, the larva is strongly "adultated" (see p. 218).

Before considering in detail these recent adult characters I must point out that an important ancient adult character can also be established within the *Mollusca*; this is a longitudinal ventral band of cilia. For example *Ischnochiton* in the Polyplacophora (Fig. 30A) and *Philine* and *Archidoris* in the opisthobranchians display this feature. According to Thompson's (1959a) and my own observations this band can be clearly distinguished medially on the foot even after the latter has become generally ciliated (Fig. 30B). (About the signification of this ancient adult character, see p. 244).

Originally the molluscs, or more correctly their ancestral forms, were devoid of a shell both in the larval and the adult stage, and even now this structure is only feebly developed or absent in the swimming larva of the Polyplacophora. Although lecithotrophic, this larva is a simple *trochophora*, i.e. without velar lobes. The larva of the "primeval" mollusc must have been of the same type, although it was certainly planktotrophic (see below).

In other words, we have reason to believe that, when shell, mantle, foot, and other adult characters appeared in the evolutionary line leading to the molluscs, this happened in the benthic phase of the life cycle, as is the case also with the adult characters in other groups (see p. 219 *et seq.*). Later on in the course of evolution adult pressure had the effect of shifting the initial formation of these characters to the pelagic larva or even to a stage of embryogenesis. This shifting to an ever earlier phase of the life cycle ("acceleration") may have taken place in several parallel evolutionary lines (see below).

The first-formed part of the shell is often distinguished by a different structure, and for this reason is usually contrasted with the later-formed main part. Consequently it is called "larval shell", but according to the argument above even this first part of the shell is nevertheless an adult character. (The special *structure* often exhibited by this oldest part might, however, perhaps be interpreted as a later evolved larval character.)

The transfer of the relatively heavy shell to the larva made it increasingly difficult for the larva to maintain itself in suspension. The rings of cilia of the *trochophora* were no longer sufficient, especially when a more important growth took place during the pelagic life. The weight was compensated by the extension into lobes of the lateral parts of the oral zone with its ciliation (prototroch, metatroch, and adoral cilia), i.e. by the evolution of the velum and thereby also of the *veliger*-larva (cf. Garstang, 1928a). There are certain indications that the process has taken place independently within Gastropoda and Lamellibranchiata (see p. 140).

This interpretation agrees well with the circumstance that in small *veliger*-larvae the velum is as a rule less well-developed, and is smaller relative to the size of the larva, than in the larger ones. Using material from Indo-Chinese waters Dawydoff (1940b) has described some giant larvae with an enormous velar apparatus (Fig. 31), and I have myself found similar forms in the Florida current near Bimini.

Thus in gastropods and the majority of bivalves evolution has on the one hand led to considerable adultation of the pelagic larva resulting in almost direct development, while on the other hand (since the pelagic mode of life was retained) one of the transferred adult characters (the shell) in its turn brought about a new, special divergence between the primary larva and the adult by causing the formation of the velar apparatus.

It is interesting to note that such large velar lobes, in common with other prominently-developed larval parts in many other groups of animals, are not successively reduced by histolysis, but are shed at metamorphosis, either as a single piece (e.g. in certain species of *Nassarius*) or as tiny fragments (for instance in the ship-worm *Bankia*; Scheltema, 1962).

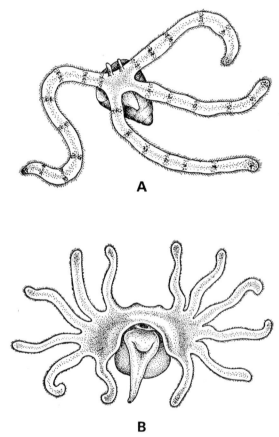

Fɪɢ. 31. Examples of giant *veliger*-larvae of prosobranchians. A, more "normal" type with 4 velar lobes; up to 16 mm between velar points. B, form with 12 lobes; 6 mm between velar points. (After Dawydoff, 1940b.)

In the latter and other forms these velar fragments may even be eaten. In one instance I have myself observed the heteropod *Pterotrachea* swallowing its large velar lobes. The specimen in question simply started to eat them while they were still in position (cf. p. 217).

It must be stressed that the velum serves not only as an organ for floating and swimming, but also for feeding. In the same way as in entoprocts and certain annelids (see pp. 108 and 159) the powerful cilia of the proto-troch drive the particles downwards to an adoral ciliated groove which is situated between prototroch and metatroch. The cilia of this groove then transport the particles to the mouth. The main traits of this phenomenon have been known for a long time and in the larva of *Crepidula* have been subjected to detailed investigations by Werner (1955). I have myself established the presence of this very efficient mechanism, which is always in action during swimming, in all true *veliger*-larvae which I have been

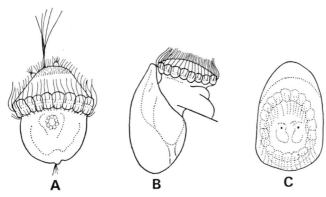

FIG. 32. Larval stages of *Acmaea testudinalis*. A, young *trochophora* (before torsion), ventral view. B, larva immediately after torsion, seen from the right. C, somewhat older larva, retracted into the shell, dorsal view. (After Kessel, 1964.)

able to study; these include the larvae of a number of prosobranchiates, of some nudibranchiates, of a gymnosomatous pteropod, and of several bivalves.

The mechanism probably occurs in all planktotrophic larvae of molluscs, and because an identical mechanism is found in other phyla (entoprocts and annelids) there is reason to assume that this character is original not only for all molluscs, but may be even older than them (see further on, p. 159 *et seq.*).

In this connection it may be noted that the otherwise primitive group of gastropods, Diotocardia (= Archaeogastropoda), possesses larvae which have undergone some changes in the ciliation of the mouth region. There is no longer an adoral ciliated groove (Fig. 32), and for this reason it is difficult to understand how ingestion of food could take place. The relatively brief pelagic phase of the life cycle and the absence of the ciliated groove lead me to the assumption that these larvae are not planktotrophic in spite of the fact that the literature contains statements about feeding. This is said to be the case in *Patella*, according to Dodd (1957), and in *Acmaea* according to Kessel (1964). However, no direct observations of ingestion seem to exist, and for this reason the assertions are open to question, especially since Anderson (1965a) has established lecithotrophy throughout the entire pelagic phase in representatives of three genera within the family *Acmaeidae*. The same applies to a species within *Patellidae* (Anderson, 1962). Further investigations would be desirable.

It should be stressed that these larvae and also those of other diotocardians possess no lateral velar lobes. The structure called a velum here is nothing more than a somewhat raised circular structure on which the prototroch is situated. Velar lobes are probably not a character which is original in all molluscs (for further details, see below). The original condition must be a "normal" *trochophora* with adoral ciliary apparatus, but the question still remains whether or not the lack of velar lobes in the larvae mentioned is to be considered an original feature. On the basis of our present knowledge the question cannot be answered. A complication is introduced by the fact that the intracapsular larva of *Theodoxus* differs from the type described by possessing lateral velar lobes which function in the ingestion of abortive eggs. These lobes must be relics from an earlier pelagic life, but it is too early to express an opinion on the problem of whether they represent an independent formation or have a common origin with the velar lobes of the monotocardians (cf. p. 140 *et seq.*).

Among the gastropods, it is not infrequent for the shell (and operculum) to be lost in the adult. In certain cases, as in parasitic (e.g. *Entoconcha*) or holopelagic forms (e.g. *Pterotrachea* and gymnosomatous pteropods), the loss is certainly connected with the new mode of life. In other cases such a connection cannot be demonstrated with certainty. Thus the nudibranchians still live in more or less the same way as their more primitive relatives with shells.

Whatever might be the cause of the loss of the shell structures, it is nevertheless an interesting fact that in quite a number of taxa of different rank this adult character now exists only in the pelagic larva. This provides an example of a course of evolution which is by no means unusual within the metazoans. By the shifting of adult characters to the larva the original divergence between larva and adult has to some extent been eliminated.

Later changes in the adult, in this case the loss of the shell, have caused a new divergence resulting in the reverse condition of a pelagic larva that exhibits adult (benthic) characters which are missing in the benthic adult.

Because the adult in the nudibranchiates is devoid of shell, operculum, and usually also of mantle fold these characters in the larvae must be designated as ancient adult characters. In this case there is no shadow of doubt that the adult of the ancestral form had a shell, etc. for the simple reason that in the more primitive relatives the adult still possesses these structures. On the basis of the principal ideas in this book, however, we should interpret conditions in exactly the same way even if all shell-bearing gastropods had become extinct without leaving any traces—an experiment in reasoning which can throw light on attempts to unravel the evolution within other groups, e.g. Bryozoa (see p. 41).

In textbooks and handbooks it is often pointed out that even those marine gastropods which in the adult stage lack a shell possess larvae which do have a shell. This rule, however, is not without exceptions. The peculiar genus *Rhodope* is without a shell throughout its life cycle (see below), and among gymnosomatous pteropods examples are known of larvae which have already lost a small remnant of the shell before they leave the envelope surrounding the eggs (see p. 133). Still more informative is the situation in the nudibranchiate *Okadaia*, where according to Baba (1937) no trace of shell, shell gland, or operculum is found. Apart from the existence of small remnants of the velum the development is direct.

It is by no means unexpected that in larvae of adults without shell, operculum, etc. these structures have disappeared or are on the way towards disappearance. Whereas ''adult pressure'' once caused a shift of the development of the shell to the larva, this pressure has, after the disappearance of the shell in the adult, begun to act in the opposite direction (cf. e.g. the disappearance of the foot in an entoproct larva, p. 106).

In the gastropods the torsion is an exceedingly important character. It occurs both in the larva and in the adult, and the question now arises as to where in the life cycle its phylogenetic appearance took place. Garstang (1928a) has discussed the problem in detail and has arrived at the opinion that the torsion made its first appearance in the *veliger*-larva. His arguments are, firstly, that the torsion should be more advantageous to the larva than to the adult, since it is necessary for efficient protection of the head (torsion is a condition for withdrawal of the head into the shell), and the second argument is the absence of fossils exhibiting the change.

I cannot enter into this interesting problem in greater detail here, but must point out that the consequences of Garstang's concept, which has found a number of adherents, are most remarkable. He considers the gastropods to be the outcome of "a sudden jump in the evolution of the Veliger larva. With its visceral dome reversed this new larva settled down and grew to maturity. . . . When its growth finished, the first Gastropod had been created."

I for my part do not regard such sudden evolution as probable. It seems obvious to me that the advantages of protecting the head cannot have been less in the adult. The ability to withdraw this part of the body, on the contrary, must be of less importance for the larva than for the adult, since the small planktonic creatures are as a rule swallowed whole. There is thus no valid reason for assuming that the torsion is an adaptation to the pelagic life, and for this reason I am inclined to believe that the phylogenetic appearance of the phenomenon occurred gradually in the adult phase to be shifted later under the influence of adult pressure, like e.g. the shell, to the pelagic larva. (The fact that the process of torsion could not be followed in the available fossil material is not a sufficient reason for the acceptance of Garstang's idea of a macro-mutation.)

Considerable variations in the life cycle are found within the gastropods. There are represented all transitions from the original condition common to all metazoans (see p. 240), with eggs that are discharged freely into the water and with a pelagic larva, to direct development of a kind in which no trace of a velum or of other pelagically adapted larval characters can be found. (Often the term direct development is applied also in cases where the change has not progressed quite so far, provided only that the young, when starting its free life has had its velum reduced and thus assumed similarity with the adult.)

An important step in the evolution took place with the deposition of the eggs in protective envelopes, at first presumably gelatinous masses of secretion, later on more complicated cocoons of different kinds. Even in certain diotocardians, e.g. *Calliostoma* (Lebour, 1937) and *Acmaea* (Kessel, 1964), eggs occur which are in some way held together.

This enclosure of the eggs was not immediately followed by any major change in ontogeny. In the first phylogenetic stages the embryos probably left the envelope at a very early stage. In the great majority of the living marine gastropods the entire embryogenesis up to the fully developed *veliger* takes place inside the envelope, and the larva starts its planktotrophic life immediately after its liberation. This mode of life continues up to metamorphosis. Frequently the larva swims about inside the cocoon before hatching.

The storage of an abundance of yolk in the egg or of special nutriment in the cocoons (albuminous secretion, abortive eggs) has allowed the entire larval phase of the life cycle to take place inside the envelope in many evolutionary lines. Planktotrophy became unnecessary and with it also the pelagic mode of life. The foremost morphological change was that the velum began to be reduced. However, such a process takes a considerable time, and for this reason there still exist even at the present time numerous holobenthic life cycles in which the velum is retained as a more or less reduced superfluous organ (unless it has in exceptional cases been put to use during life in the cocoon; see p. 127).

We notice with interest that even within the Pulmonata there is a wide range of variation in the life cycle. Even in forms which are usually referred to the Stylommatophora there is not only a *veliger* stage which is enclosed in the egg, but also pelagic larvae. According to Thorson (1940) a pelagic larva occurs in the genus *Onchidium*. (According to Fretter (1943) *Onchidium* is, however, an opisthobranchian.) It is most interesting to find that in the pulmonates too, cases of planktotrophy are still found. This applies, according to Anderson (1965b), to a species of *Siphonaria*.

Understandably enough the ontogeny is strongly transformed in the terrestrial forms of Stylommatophora. It might seem more surprising that even in the marine genus *Rhodope* which according to Riedl (1959, 1960) probably belongs to the stylommatophores, the ontogeny has changed to direct development. *Rhodope* is, however, strongly altered even as an adult. (It is significant that it once was interpreted as a turbellarian.) This aberration, whatever its cause, has in its turn exerted a strong influence on the embryonic stages, and as far as we can see this is again a result of the adult pressure. Thus a delimited foot, mantle fold, or shell are no longer developed (but nevertheless an indication of a shell gland) and already at hatching the whole surface of the body of the juvenile is covered with cilia exactly as in the adult. It is not very surprising that in such an altered ontogeny the velum can be traced only as unimportant "Velarwülste", but this is nevertheless of value as an indication that a *veliger* formerly existed also in the evolutionary line leading to *Rhodope*.

To me it appears obvious that in *Rhodope* as in all other cases in which an indication of a velum can be traced in the course of the embryogenesis (and in other cases, too, where this is no longer feasible) evolution has progressed from a free *veliger* towards direct development and not in the opposite direction. This view is based not only on my general ideas about the pelagic larvae, but also on the fact that I consider it absurd for an organ which represents an adaptation to pelagic life to evolve in the enclosed embryogenesis of a life cycle with direct development. This is, however, the opinion held by Riedl, and for this reason I shall return to the subject later (p. 143).

Differences in ontogeny between closely related forms such as species of the same genus are not rare. Thus in *Siphonaria* (Basommatophora) the species *S. sipho* has a pelagic *veliger*, while *S. kurracheensis* has a holobenthic life cycle (Thorson, 1940). In certain cases of this kind the difference may be connected with the occupation of separate biotopes. The temperature of the environment is also known to exert an influence, direct development being coupled with cold water, and a pelagic larva with warm (Thorson, 1940). Yet the differences cannot always be explained by the environmental conditions. Thus according to Kohn (1960, 1961a and b) the majority of the tropical species of the genus *Conus* has pelagic larvae while some of them pass the whole or almost the whole *veliger* period enclosed in the envelope of the egg, in which case the juveniles start their creeping life immediately after hatching. In the latter category the eggs are, as is generally the case, bigger and richer in yolk. It is evident that this elimination of a pelagic mode of life in *Conus* as well as in *Siphonaria* has taken place *within* the genus in question, perhaps even independently in several species.

According to information supplied by Rasmussen (1956) *Rissoa membranacea, Brachystomia rissoides,* and *Embletonia pallida* are species which are, so to speak, part way between a pelago-benthic and a holobenthic life cycle. These species exhibit within one and the same area both types of life cycle. (For other species with variable life cycle, see Fioroni, 1966, Table 48.)

Recently Thompson (1967) has prepared a compilation of the information available on the life cycle in Opisthobranchia. He distinguishes three types—those with:

1. planktotrophic larva,
2. lecithotrophic larva,
3. benthic direct development.

In the last type *veliger* structures can be traced more or less distinctly in the course of embryogenesis.

All this agrees perfectly with my general concept. Lecithotrophy and direct development have obviously originated independently several times in the molluscs.

It is not possible here to enter more closely into the stages in the reduction of the velum or the different types of intracapsular development related to the supply of albuminous secretion or abortive eggs. (Ankel, 1936, gives a good account of the prosobranchians.) We shall only deal briefly with some conditions that are of interest here.

If the development in cocoons and other types of egg envelopes as the result of the rich supply of nutriment has on the one hand been simplified by the reduction of the velum, it has on the other been complicated in several respects by adaptation to new conditions, implying among other things, the evolution of transitory embryonic organs of different kinds (cephalocyst, podocyst, dermal kidney, etc.). Changes of the latter kind have caused Portmann (1955) to speak about a "métamorphose abritée" —a hidden metamorphosis. Here, however, we are no longer concerned with a larva but with an embryo. The embryogenesis has been lengthened at the expense of the pelagic larval life.

The transfer of the whole larval phase of the life cycle to the cocoon has, however, not always removed the importance of the velum. In at least one case its feeding function has been retained in spite of the changed circumstances. This remarkable adaptation has been established by Fioroni and Sandmeier (1964) in a prosobranchian from the Mediterranean (according to Fioroni, 1966, a species of *Cassidaria*). In this species the velum is well developed. It is almost as big as in forms with a pelagic mode of life, and with its aid the larva is able to swim about in the egg capsule. The ingestion of food takes place in the same way as in pelagic larvae— particles of food (in this case the yolk grains of the abortive eggs) are taken up in the adoral ciliated groove through which they are carried to the mouth. A new feature is, however, that the abortive eggs are grasped by the reverse side of the velar lobes. In this way a pouch is formed, in which the eggs are set into rotation by the activity of cilia (of the metatroch ?). The egg membrane is destroyed by friction, and the yolk grains are gradually loosened from the periphery of the egg. Then the adoral ciliated groove, situated between the prototroch and the metatroch, effects their transport to the mouth (see Fig. 33A). It is reported that the larva can continue to swim during the feeding process. At the end of the period in the cocoon, the velum is reduced, and after release from the egg capsule the young adopts a creeping mode of locomotion.

It is still uncertain whether the conditions reported by Fioroni and Sandmeier agree with some stage in the evolutionary lines leading to other life cycles with abortive eggs. In at least some of the forms which

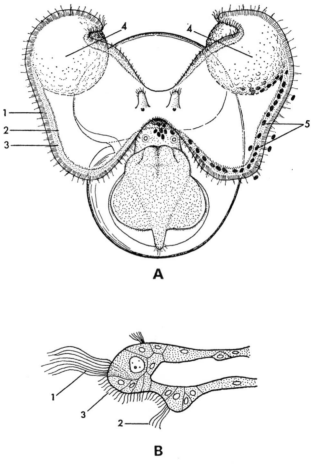

Fɪɢ. 33. A, older larva of *Cassidaria* sp. taking up yolk granules from an abortive egg in the cocoon. (After Fioroni and Sandmeier, 1964—simplified and slightly altered.) B, section through the velar margin of *Crepidula*. (After Werner, 1955.) 1, prototroch; 2, metatroch; 3, adoral ciliated groove; 4, abortive egg which has been taken up by the velar lobes; 5, loosened yolk granules on their way towards the mouth in the adoral ciliated groove.

swallow entire eggs their transport to the mouth is likewise effected by cilia (my own observations; see also Fioroni, 1966). Additional investigations into the feeding mechanism in forms with abortive eggs would be desirable. It would be interesting if traces of the original mechanism could be found.

In a viviparous prosobranch from the Fiji Islands, for which Hubendick (1952) has created the genus *Veloplacenta*, embryogenesis leads to young with a large-lobed velum lacking cilia. This velum obviously fulfils some function during the intra-uterine life. The strong growth makes it certain that food is ingested, and Hubendick considers it probable that the velum in conjunction with the wall of the uterus functions as some kind of placenta. I consider it probable that this is actually so, especially in view of the report that the large velar lobes are devoid of cilia—if this is the case it is a very remarkable result of evolution. However, for this form, too, further studies are desirable.

There are other cases among the viviparous gastropods in which the embryo is supplied with nutriment inside the uterus, such as, for example, *Paludina*. It becomes evident that this is the case from the fact that the embryo reaches a considerable size within the mother animal although the egg is exceptionally small. The small size of the egg is evidently connected with the supply of food and is certainly secondary. Originally viviparity implied no more than the transfer of a period of development to the interior of the mother animal. The more complicated conditions, like those found in *Veloplacenta*, must certainly be derived from such a simple situation.

In several instances among the Lamellibranchiata brooding and consequently direct development are found. As an example *Sphaerium* may be mentioned. A *trochophora* stage can be distinguished in embryogenesis, but no distinct traces of the velum are retained. When the juveniles leave the mother animal, they are on the whole similar to the adults.

As far as the lack of a free *veliger*-larva is concerned the unionids, too, display direct development. Here conditions are, however, complicated by the fact that the parasitic life of the juvenile has led to its alteration in a very special way. During embryogenesis it is transformed into a so-called *glochidium*-larva which immediately after hatching is ready to parasitize fish. On the basis of present knowledge it is impossible to decide whether the *glochidium* should be interpreted as a secondary larva, i.e. that it has been evolved in a life cycle with typical direct development, or whether it has originated directly from a free-swimming *veliger*-larva. Because we are dealing here with fresh-water forms the former alternative seems to be the more probable.

It has already been pointed out that in the gastropods the velum is a purely larval character. As far as is known there are no cases where it has been retained in the adult. The cases of a holopelagic life cycle which occur in a number of forms is thus not the result of neoteny in the way that the *veliger*-larva has become sexually mature and the benthic phase accordingly has been eliminated. This is rather remarkable in view of the efficiency of the velum as an organ of both locomotion and feeding, and the long duration of the pelagic phase in certain forms with benthic adults (see e.g. Fig. 31).

The foot seems to have played the decisive rôle in the transition to holopelagic life. In the majority of the holopelagic forms the foot is still functional, though transformed into a swimming organ in different ways. Originally it was a creeping organ and it still functions as such in the great majority of the gastropods. In a number of opisthobranchians, e.g. *Pseudobranchus*, *Akera*, *Aplysia*, and *Elysia*, its peripheral parts are more or less extended like fins. This enables the animal to swim upwards from the substratum, even if creeping on the bottom is still the main mode of locomotion. As far as can be judged, gradual evolution in this direction has led to the holopelagic life cycle in the ancestral forms of such types as *Limacina*, *Cavolinia*, *Clione*, etc., where the peripheral parts of the foot have been transformed into very efficient paired organs for swimming (Fig. 34E). In the heteropods, on the other hand, the course of evolution is less evident, but since even here the foot is the organ of locomotion, it appears to me most probable that in this case, too, the adult has changed gradually from benthic to pelagic life. There is at all events nothing to indicate that the adult evolved from a neotenic *veliger*-larva.

In the majority of the holopelagic forms two entirely different swimming organs function during the life cycle (in the gymnosomes there are even three; see below): first the velum and later on the foot. While the velum exists it is usually well developed. An appreciable reduction has been found only in a few cases (see below).

The shell is a structure which without doubt is influenced by the holopelagic mode of life. It is evident that the shell, and also the operculum, constitutes, as far as weight is concerned, a greater burden to the adult than to the larva. For this reason the shell of the adults has been subjected to some degree of reduction or has been lost altogether.

It is an interesting question to consider to what extent the effect of adult pressure on shell and velum can be traced in holopelagic forms. Considering first the thecosomes we find that the adult has retained its

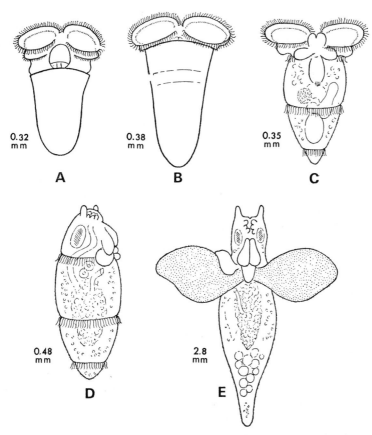

FIG. 34. *Clione limacina*. A and B, *veliger* larvae with shell. C, *veliger* larva which has lost the shell and acquired the three rings of cilia. D, larva having lost the velum. E, juvenile with growing eggs in the caudal part of the body. (After Lebour, 1931—simplified.)

shell, even if it has become thinner and, as far as the Cavoliniidae are concerned, has assumed a profoundly altered shape. As can be expected it is consequently also retained in the young. As usual it is developed early in embryogenesis.

The fins, which derive from the foot, are formed early, and are ready for use immediately at or shortly after hatching. Because the newly-hatched larva is thus able within a short time to feed in the same way as the adult there is nothing to prevent an early elimination of the velum, which consequently takes place in certain cases. Using material from the Florida current (February, 1965) I have had the opportunity to make the following observations.

In a commonly-occurring species of *Limacina* it was found that in larvae with a shell of about 350 μm in size (greatest diameter of the spiral) the velum was well developed (each lobe of approximately the same size as the shell) and served as a swimming organ. The fins were still small and virtually immobile. By the time the shell had reached the size of about 500 μm, the velum had disappeared completely, and the fins were used in the ordinary way. It is evident that in the form in question the velum persists only for a short time compared with the larvae of most benthic gastropods.

Cavolinia has progressed a step further towards direct development. In a species of which I found several "cocoons" with freshly laid eggs in plankton samples from the Florida current, hatching generally took place at night. In the morning the young exhibited no trace of the velum and swam about using their fins exactly like the adults.

I have had no opportunity for any detailed examination of development within the gelatinous cocoons, but the following observations could be made. Embryogenesis is very fast, so that by the second day after the beginning of cleavage both velum and fins are clearly distinguishable. From their emergence the cilia of the velum move vigorously inside the egg membrane. The fins are at first held motionless, but even before hatching, which in one case was found to take place at the beginning of the fifth day, they start to execute some jerky movements. In this case the larvae swam by means of the velum only for some minutes after hatching. The velum then shrank, and at the same time the fins were extended and started to perform swimming movements. These changes occurred very rapidly. After about 5 min no trace of the velum was left, and the young were in all essential respects like the adults. Unfortunately I had no opportunity to observe the way in which the velum is eliminated, but the speed of the process indicates that its cells are discarded. Afterwards they may perhaps simply be ingested (cf. above, p. 120).

Surprisingly enough the process described is not found in all thecosomes. Considerably different conditions are found, e.g. in *Creseis* in which Gegenbaur (1855) observed a large-lobed velum in an obviously quite old larva that was, however, still entirely without fins.

Evidently the thecosomes exhibit great variations over the disappearance of the velum and the development of the fins. But irrespective of the time when these changes take place, their mutual connection is always obvious, and the simultaneous function of both swimming organs is restricted to a short period.

Since these differences, though considerable, represent only different degrees in the reduction of the velum and different times for its replacement by the final swimming organ, no major taxonomic significance should be attributed to them. On the basis of our present knowledge we cannot tell why, within a uniform group, the adult pressure is so different or at all events has produced such different results in the various forms.

In the gymnosomatous pteropods the adult has no shell, and the body is markedly changed—all adaptations to the pelagic mode of life. On the other hand the larva possesses, in certain cases, a well-developed shell which is later lost. These facts are known, for example, from Lebour's (1931) investigations of *Clione limacina* (Fig. 34).

I myself have found in the Florida current a gymnosome which on several occasions laid eggs in the storage dishes (February).* At hatching the larvae were found to be without shell (Fig. 35C). An examination of the abandoned gelatinous mass showed, however, that small, almost watch-glass-like shells had been shed inside it (Fig. 35B). I consider the marked reduction of the shell and its early shedding to be the result of adult pressure which thus has had a much more pronounced effect in this form than in *Clione limacina*.

The free-swimming larva of the gymnosome described is not only without shell, but also without velum (Fig. 35C). During the fourth day of embryogenesis, which altogether takes only about five days, a remnant of the velum could nevertheless be observed (Fig. 35A). Swimming was found to be performed by means of the three rings of cilia characteristic of these larvae. The velum evidently disappears within the egg or immediately after hatching. This form can be said to exhibit almost direct development.

Other members of the gymnosome group are rather different. *Clione* has a relatively well-developed velum which according to Lebour (1931)

* I do not know to what species this gymnosome belongs. It is characterized by an opaque, greyish appearance and absence of fins. (Locomotion was effected by the rings of cilia which are common in the larval stage.) The size of the egg-laying specimens was only about 1·5 mm. It is possible that growth continues, and that fins are developed later on.

functions for over two weeks (Fig. 34A–C). In larvae taken from the plankton Gegenbaur (1855) could establish the existence of a very large velar apparatus, and in the Florida current I myself have found several specimens of a similar form (Fig. 35D). Interestingly enough the ciliation of the velum proved to have the same nature and function as in other

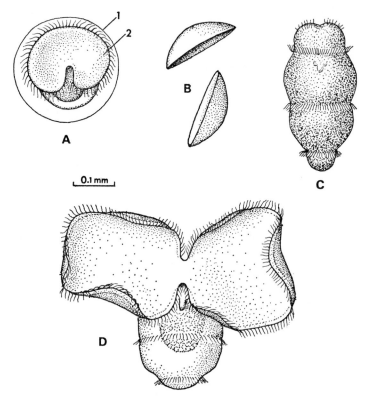

Fɪɢ. 35. Developmental stages of gymnosomes. A–C, form reared in the laboratory (see footnote on p. 133). B, shed shells. D, *veliger* of another species, taken directly from the plankton. 1, envelope of the egg; 2, velum. Magnification approximately uniform.

planktotrophic larvae of molluscs, i.e. the food particles are transported to the mouth in an adoral ciliated groove (see above, p. 121). A specimen of the form D in Fig. 35 metamorphosed in the course of one night. In the morning no traces of the velum were left, and the shape of the young was essentially the same as that of the form in Fig. 35C. As far as one can guess the velum had been ingested (cf. p. 120).

As can be seen there is an enormous difference between the two forms A–C and D in Fig. 35 with regard to the velum. The great reduction of this organ (and also the shell) of the former species I consider to be a result of adult pressure. There must be no doubt that there has been evolution within the gymnosomes towards direct development, even if the rate has been different in different forms (for the probable explanation of this evolution, see p. 236).

It is remarkable that in the period of the life cycle of the gymnosomes that precedes the full development of fins a third type of organ of loco-motion has been intercalated, viz. the three rings of cilia characteristic of the group. In *Clione* (Fig. 34C and D) this is particularly evident. It is obvious that these rings are new formations in the larva which have arisen after the elimination of the shell in the adult, and which thus constitute a character specific to the gymnosomes. Ontogenetically the rings are formed only after the larva has dropped the shell (otherwise at least the two posterior rings would be covered by the shell). These rings thus characterize a second larval stage. (N.B. this is not equivalent to a "secondary larva", see p. 8.) The rings are retained for a remarkably long time, the posterior ones still being found in certain forms, when the length of the body has reached more than 2–3 mm. In the form which laid eggs in the dishes the rings may be found throughout life, and this would then constitute a kind of neoteny. (N.B. the fins of the species are still not developed when sexual maturity is reached.)

However, I find no reason to assume that the gymnosome group as such originated neotenically, i.e. was derived from some sexually mature larva, provided with three rings, of some benthic ancestral form. It is probable that the ancestral form of the gymnosomes became holopelagic in the same way as that of the thecosomes, by a gradual transition of the origin-ally benthic adult to the pelagic zone. This might have been effected by increasingly protracted excursions from the bottom. There is at least no evidence against such a course of evolution.

The existence in the pelagic zone of the relatively heavy adult of the gymnosome ancestor, which probably carried a shell, was dependent on the evolution of the fins. I therefore consider it most probable that the elimination of the shell of the adult first took place after the adoption of holopelagic life. The process would thus have been the same as in the *Heteropoda*, where the genera *Atlanta*, *Carinaria*, *Pterotrachea* represent a series of types in an evolutionary process of this kind.

The three rings of cilia are in any case young formations, which have

arisen in the special line of evolution that leads to the present gymnosomes. It is difficult to tell why in contrast to other groups this third kind
of organ of locomotion should have been inserted in the life cycle.
Secondary rings of cilia also occur, however, in the older pelagic larvae of
other groups of animals, e.g. the *doliolaria* of the echinoderms and poly-
trochous larvae of annelids, and they are formed on different parts of the
body.

The above makes untenable the opinion of Söderström (1925b) who
asserts that the cilia-rings of the gymnosomes larvae are homologous with
a metamerically repeated prototroch, and thus points to an origin of the
molluscs from the annelids.

We have now dealt with life cycles that contain or have contained a
larva of the common *trochophora* or *veliger* type. There is, however, within
the molluscs another very special type which cannot be omitted here. For
reasons which will become evident later it may be called the "test-
type". It occurs in Solenogastres and among the bivalves in Protobranchiata. I shall begin with the larva in the latter group.

This larva, which has been described by Drew (1899) in the genera
Yoldia and *Nucula*, is characterized in its external aspect mainly by the
remarkably large cells of the surface ectoderm. These cells are arranged in
five circles of which the three in the middle each bear a ring of cilia (Fig.
36). In front a long flagellum is found which is composed of cilia or sensory
hairs. The posterior end contains an opening which is interpreted by
Drew as blastopore. This interpretation can hardly be correct, at least in
the older stages (cf. Fig. 38A). The surface ectoderm, the test envelope,
is lost at metamorphosis, and is said to be simply dropped (before this
occurs the final ectoderm has been formed by immigration of certain cells
from the surface layer). It is important to note that the larva is lecithotrophic.

Drew compares the test envelope with the velum of the larvae of the
other molluscs and supposes that they are homologous structures—a concept which later appeared repeatedly in the literature—and that the test
larva belongs to an original type from which the other molluscan larvae
evolved.

I cannot share this opinion. Even if much still remains to be elucidated
about the larvae of the protobranchians, I consider it as evident that
they are not an original form but are considerably altered (for further details see below). It is, however, of interest that there are certain primitive

features left: the sexual products are freely discharged into the water as is usually the case in bivalves, fertilization is external, and even early embryonic stages swim about by means of their cilia.

The above applies to *Yoldia limatula* and *Nucula proxima*. In *Nucula delphinodonta* conditions are altered still more. The eggs, which are

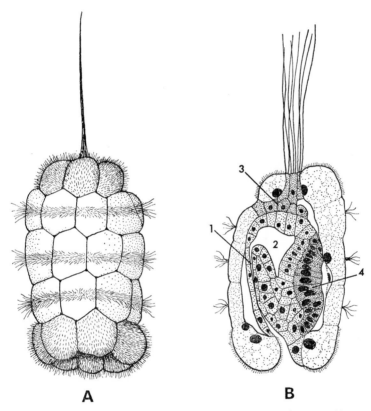

FIG. 36. *Yoldia limatula.* A, larva in lateral view. B, median sagittal section. Test envelope finely dotted. 1, stomodaeum; 2, mesenteron; 3, rudiment of cerebral ganglion; 4, shell gland. (After Drew, 1899.)

relatively large and rich in yolk, are deposited in a cocoon that is fixed to the posterior end of the mother animal and is in open communication with the mantle cavity. In this cocoon the whole development right through to a test larva takes place. The test envelope is shed as in *Yoldia*.

Certain larvae in Solenogastres exhibit a remarkable similarity with the larval type of the protobranchians. Pruvot (1890, 1892), Heath (1918), and recently Thompson (1960), who has carried out detailed examinations of *Neomenia*, have all established the occurrence of a test envelope, inside which much of the development takes place (Figs. 37 and 38). It cannot be denied that this temporary ectoderm in the larva of *Neomenia* constitutes a great similarity with that of the larvae of *Yoldia* and *Nucula*, in

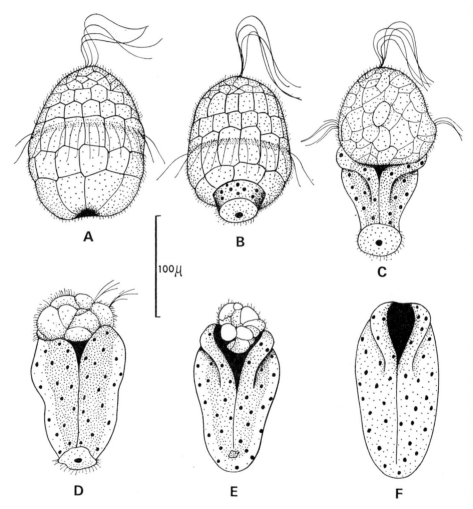

FIG. 37. Developmental stages of *Neomenia carinata* in ventral view. A, newly hatched larva. The remainder of the series shows how the test envelope extends forward and is incorporated into the body. The larger black dots represent calcareous spicules in the final epidermis. (After Thompson, 1960.)

fact so great that Thompson's suggestion "that the common ancestors of all the Mollusca may have had a larva of a similar type" may be quite plausible in the light of present ideas on larval forms.

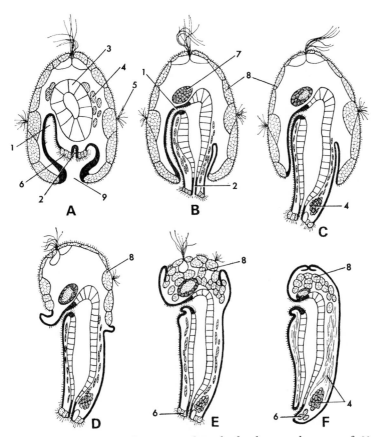

FIG. 38. Diagrammatic sagittal sections through developmental stages of *Neomenia carinata*. The stages shown here correspond roughly to those in Fig. 37 with the same designations. Test envelope dotted; final ectoderm black. 1, stomodaeum; 2, proctodaeum; 3, mesenteron; 4, mesoderm; 5, prototroch; 6, telotroch; 7, brain; 8, test envelope (note how it is incorporated into the anterior end of the body: D–F); 9, pseudoblastopore. (After Thompson, 1960.)

However, from what I have said above it must already be apparent that I do not consider this interpretation as correct. Some critical remarks are necessary since Thompson finds that his results lead to the following conclusion (p. 277): "In the present state of our knowledge the Aplaco-phora might be considered to be, among the Mollusca, most closely allied to the primitive Lamellibranchia, since the development of *Neomenia* shows such striking similarities to that of the Protobranchia."

To me it seems quite obvious that the test type, wherever it occurs, is an altered type that has arisen from an ordinary, though lecithotrophic *trochophora*. The first stage of the divergence, at least in *Neomenia*, con-sists of an enlargement and folding backwards of the preoral part of the body wall. In this way the oral and postoral parts have become covered (Fig. 38A and B). Because the larva was lecithotrophic, this could happen without disadvantageous consequences. The rings of cilia in the oral zone had already been altered in such a way that they served exclusively for locomotion.

If the test type were original for the Mollusca, this would imply that the ordinary *trochophora* which is found in different instances within the group, and with it also the *veliger*, has arisen from the test type, and that the agreement of the *trochophora* of the molluscs with that of the Annelida and other *trochophora* groups is a mere convergence. Already this makes Thompson's conception impossible. To this it can be added, that the test larvae are lecithotrophic, as has already been mentioned. According to my earlier conclusions the primeval molluscan larva was planktotrophic.

How could it then happen that in two widely separated cases within the Mollusca the ordinary *trochophora* could be altered into a test larva, while in other cases the changes resulted in a *veliger*? I have already arrived at the conclusion above that the formation of the velum was a reaction to the shifting of the inital development of the shell to the pelagic phase. (During and after the emergence of the velum the original method of feeding was retained.) As is so often the case with adult characters this shifting of the shell to the larva has taken place independently in several instances, and thereby the emergence of the velum has, so to speak, been provoked in different lines of evolution leading to the recent molluscs. Thus it is by no means necessary for the ancestral form common to Gastropoda and Lamellibranchiata to have had a *veliger* (but certainly an ordinary *trochophora*), and since at the present time the protobranchiate bivalves possess such an aberrant larval type, which may be derived from an ordinary *trochophora*, but not from a *veliger*, it seems that the velum originated in the "higher" mussels *after* the splitting-off of the evolu-tionary line leading to the Protobranchiata. In this latter line the *trocho-phora* became lecithotrophic and was altered into a test larva. The absence of a true velum in primitive gastropods (here we find only a raised circular prototroch, see Fig. 32) may be additional evidence for the idea that the velum of gastropods and bivalves has arisen independently in these groups (cf. also above, p. 122, concerning *Theodoxus*).

Thus we can explain the emergence of the *veliger*-larva satisfactorily; it is comprehensible that the increased strain to which the locomotive ciliation became exposed by the shifting of the shell to the original larva, i.e. the *trochophora*, has caused the evolution of the velum, and also that this happened independently in different instances. In each case identical strains, caused by the same factor, acted on the same larval type in one and the same group. For this reason the results everywhere were essentially identical. It should be kept in mind that the change into the *veliger*-larva took place with retention of planktotrophy and of the original feeding mechanism (see p. 121).

It is much more difficult to express an opinion about the cause behind the emergence of the test type in two different instances. However, we must not forget that the evolution of a test envelope is not exclusive to the molluscs. The so-called serosa in the embryo of *Sipunculus* is a fold of the same kind (see p. 147 and Fig. 40A), but the serosa, too, is phylogenetically an isolated structure. In my opinion such folds have thus evolved independently in Sipunculida, Solenogastres, and Protobranchiata.

The question also arises, whether the similarity between the test larva of the two last-named groups has not been overrated. As mentioned before, in *Neomenia* it is a preoral zone that has been enlarged and folded backwards over the mouth and the postoral parts of the body. In the protobranchians, on the other hand, as far as can be judged from Drew's investigations, even the oral zone seems to form part of the backwardly folded portion. By the folding the stomodaeum has been extended in the shape of a long tube. Other differences, too, exist, one in particular concerning the final fate of the test envelope. While in the protobranchians this envelope is simply shed, in *Neomenia* it is incorporated into the body in such a way that the final ectoderm of the anterior end grows around it. This remarkable process is represented in Figs. 38D–F.

According to the idea proposed here there is no direct phylogenetic connection between the test type and the *veliger* type, but both are derived independently from an ordinary *trochophora*. This concept would not necessarily be upset even if by any chance a typical *veliger* were to be found in some protobranchiate. This would not necessarily imply any more than that the original larva, the *trochophora* (under the influence of the acquisition of the shell and a lengthy pelagic life), would have evolved velar lobes in yet another instance (cf. also the *rostraria*-larva, p. 161).

In this connection it might be asked, why the test type could not have originated from the *veliger* type instead of from an ordinary *trochophora*. In this case it would not be necessary to assume that velar lobes have arisen independently in two or more instances. I cannot agree with such an interpretation. The original larva of the Mollusca and also of the Annelida and other groups is a simple *trochophora*, i.e. a larva in which the lateral portions of the mouth zone are not extended as lobes. Like the zone of the prototroch in the original *trochophora* the test envelope is a circular structure, and nothing in its organization indicates that it can be derived from paired lateral lobes serving for feeding. It is much more logical to compare the test type with a lecithotrophic larva such as that in *Dentalium*. This comparison, actually made already by Drew, is more productive, and it is very probable that the test type is derived from just such a type. The larva of *Dentalium* is not a genuine *veliger*. Its ''velum'' is a circular, purely preoral structure, thus without metatroch and without any function in feeding. Like certain larvae of diotocardians (see p. 121) this lecithotrophic larva is most properly designated as a *trochophora*. The term *veliger* I want to restrict to larvae in which the lateral portions of the mouth zone, the metatroch included, have been extended into paired lobes.

In summing up I should like to make the following remarks about the evolution of the molluscan larvae. Here, as in other *trochophora* groups, the primitive condition is a planktotrophic *trochophora*. This has been altered in the following way. In the Polyplacophora (all of them ?) and in certain Solenogastres (e.g. *Epimenia*) it has become lecithotrophic, but has, apart from the reduction of the adoral ciliary apparatus, probably not undergone any major changes. The same applies to the Scaphopoda. In some Solenogastres (e.g. *Neomenia*) and in protobranchiate bivalves lecithotrophy has also arisen independently, and after this it was essentially a preoral zone with the prototroch that was transformed into a test envelope. In the other bivalves and in at least the majority of the gastropods the still planktotrophic *trochophora* was transformed into a *veliger*-larva, provided with a velum that performed both locomotive and trophic functions. This change was the result of the shifting of the initial development of the shell to the pelagic phase. Finally, in many different evolutionary lines ontogeny is altered towards direct development. In certain cases the primary larva has even been completely replaced by direct development. The existence of a *veliger* in the evolutionary line leading to the *Cephalopoda* cannot be directly proved, but it appears by no mean improbable on account of the occurrence of shell in this group.

My discussion of the molluscs might have terminated here, had not Riedl (1960) in his interesting study of *Rhodope* expressed a general idea about the evolution of the life cycle in the group, which cannot be left unopposed.

According to Riedl *Rhodope* exhibits direct development. This statement can be accepted as correct in spite of the fact that for a couple of days after the hatching the lecithotrophic ''larva'' moves about swimming freely, presumably in close proximity to the biotope of the adult. Thus no real pelagic phase seems to exist. Swimming is performed by means of a general ciliation of the same nature as in the adult. During this free period no velum is found, but earlier in the development unmistakable ''Velarwülste'' could be observed inside the envelope of the egg. There is consequently no doubt that a *veliger* stage, although considerably altered, forms a part of embryogenesis. I must stress this point in spite of Riedl's attempt to deny the occurrence of both *veliger* and *trochophora* stages. Instead of these Riedl distinguishes in the embryonic development a ''*Reisinger*-Stadium''. This he defines as a phase of development, ''welcher, unabhängig vom Trochophora-Stadium, dieses gewissermassen ersetzend, im Falle indirekter Entwicklung zum Veliger-Stadium führt, im Falle direkter Entwicklung selbst den Höhepunkt der Ausprägung der Larvalmerkmale darstellt'' (1960, p. 251).

The formation of this *Reisinger* stage which should by no means be restricted to *Rhodope*, but should be applied to almost the entire complex of molluscs, starts immediately after gastrulation. Its most prominent characters are the following: shell gland, ventro-frontally situated mouth opening, velar swellings, more or less widely distributed ciliation at the anterior end, statocysts, and frequently also protonephridia. The stage is completed after the development of the organs ''indem dann die Velarwülste entweder gleich verstreichen oder sich noch zu Velarlappen vergössern, um erst nach einer pelagischen Periode zu verschwinden''.

Riedl himself establishes the existence of transitions between the *Reisinger* type and the *trochophora* type within almost all groups of molluscs and stresses that the relationship is unmistakable. When introducing a new concept, however, Riedl seems to do this because he regards direct development as the type of ontogeny which is original for the molluscs. Consequently pelagic *trochophora* and *veliger* would be secondary forms, derived from the *Reisinger* stage. Riedl even goes so far as to suggest that on different occasions this stage has given rise to a pelagic larva.

Riedl gives no valid evidence, however, for his concept. The characters described for the *Reisinger* stage are not sufficient reasons, neither is the fact that several forms with a pelagic *veliger* (e.g. *Crepidula*) in their embryogenesis exhibit a stage resembling the *Reisinger* stage in forms with direct development. In my opinion the reason for this similarity is that in either case embryogenesis has been lengthened; the *trochophora* has become an embryonic stage that has been altered in a similar way.

By his presentation Riedl joins the authors who on untenable grounds oppose the current and well-founded concept that the *trochophora* is a very ancient, original larval type, common to several phyla (see also p. 250).

It is, however, of interest to see that Riedl is obviously not quite sure of his own interpretation. At the end he makes the following concession (p. 257): "Letzten Endes hängt die Frage aber mit der phylogenetischen Beurteilung der Planktonlarven überhaupt zusammen." This is a perfectly correct point of view. It is exactly a "Beurteilung" of this kind which I have attempted in this book. In doing so I have arrived at the conclusion that direct development cannot possibly be the original state of affairs. The situation in *Rhodope*, as in all forms with a "*Reisinger* stage", must be interpreted in a way contrary to Riedl's idea. The type of life cycle which is original for the molluscs included a pelagic *trochophora*. In certain evolutionary lines a similarly pelagic *veliger* followed later. As I have already pointed out, the velar lobes are adaptations to the pelagic mode of life, and it cannot possibly be conceived that they made their first appearance in a type of ontogeny involving direct development. The velar swellings which appear during the embryogenesis of *Rhodope* and other forms are relics of ancient velar lobes which in the past functioned in a pelagic larva. I shall once more stress the transformation. The *trochophora* and *veliger* stages have been depressed into embryogenesis and have been more or less altered, in exceptional cases to such a degree that the larval characters were altogether lost. There is no reason to accept the proposed term "Reisinger stage". It would have been justified if direct development had represented the original conditions in the molluscs. There is now a danger that the term will be a source of confusion.

Under these circumstances it is remarkable that such an experienced scientist as Portmann has not been left unimpressed by Riedl's presentation. In his introduction to the molluscs in "Traité de Zoologie" (Tome V, fasc. II, p. 1643) he writes: "Il se peut que l'ontogenèse des premiers Mollusques ait débuté par un développement direct, sans stade larvaire intercalé", even if he does cautiously add: "La signification évolutive de ce 'stade de Reisinger' reste a déterminer." Fioroni (1966), on the other hand, is of the opinion that in the molluscs indirect development with a free larval phase is more primitive than direct development. Personally I consider that few phylogenetic questions are so relatively easy to determine as the interpretation of the two main ontogenetic types that occur in this group.

Sipunculida

The general rule of a pelago-benthic life cycle applies also to the sipunculids. There exist no exceptions for the main biotope of the adult, and it might be added that all species are sedentary (in part burrowing). In the juvenile phase one exception is, however, known: the small hermaphroditic species *Golfingia minuta* has changed to a holobenthic life cycle (see below).

The sipunculids are also primitive in their method of liberation of the gametes and the mode of fertilization. As far as is known both kinds of sexual cells are discharged freely into the water, and fertilization is external. (In *Golfingia minuta*, however, the eggs remain, in the tube of the mother animal, where a primitive type of brooding thus takes place.)

Among the larvae of the sipunculids two types can be distinguished; one type has a lengthy pelagic and planktotrophic life, the other is lecithotrophic, only swimming freely for a short time—in *G. minuta* this does not occur at all. For a correct understanding of the phylogeny of the larvae these two types and their ontogeny must be considered separately.

The planktotrophic type which in the literature is generally called "*pelagosphaera*" has only quite recently become satisfactorily known as far as morphology and behaviour are concerned (Jägersten, 1963).* The most striking feature of its organization lies in the existence of a distinctly

---

* Häcker's (1898a) different forms of "baccaria" are nothing more than larvae of sipunculids, as Häcker himself suspects. However, his account of them has accidentally come to include a planula of an actiniarian.

developed head and, ventrally behind the mouth, a ciliated creeping organ in the shape of a lip-like lobe (Fig. 39A and C).

The larva of *Sipunculus nudus*, the development of which has been studied mainly by Hatschek (1884), belongs to this planktotrophic type, although it is not designated as *pelagosphaera*. Hatschek's investigations show that

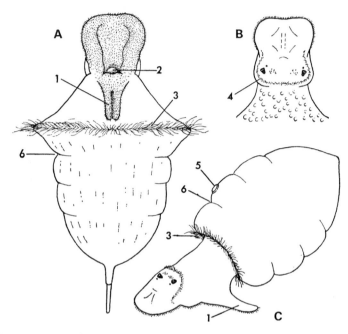

FIG. 39. *Pelagosphaera* from the Florida current (probably a species of *Phascolosoma*). Length of body approximately 1 mm. A, swimming in ventral view, with anterior end (introvert) and "tail" extended. B, head in dorsal view. C, individual creeping on solid substratum. 1, creeping lobe; 2, pharyngeal bulb; 3, metatroch; 4, prototroch; 5, anus; 6, boundary between introvert and abdomen. (After Jägersten, 1963.)

the larva is an unmistakable *trochophora* with an early developmental phase which has been altered in a rather special way.

The early development of *Sipunculus nudus* (and probably all other forms with a typical *pelagosphaera*) takes place inside the egg membrane. A particularly characteristic feature of these young stages is the very strong development of the zone of the prototroch. In this zone the ciliated epithelium has been folded as shown in Fig. 40A. The folding takes place in a forward as well as a backward direction with the result that all or practically all of the body of the embryo is enveloped by a double epithelial

membrane ("serosa"). The outer layer thus consists of the flattened cells
of the prototroch, from which the cilia penetrate the egg membrane and
permit active swimming even in the very young embryo. This hypertrophied prototroch is fully developed and functional even before the end
of the gastrulation. During this first phase of development feeding is

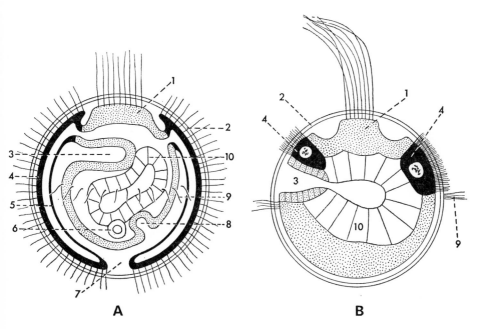

A   **B**

Fig. 40. Diagram showing the organization of the free-swimming embryo of *Sipunculus
nudus*. Construction after Hatschek's (1884) figures. Serosa black; final ectoderm dotted.
1, apical disc; 2, egg membrane; 3, stomodaeum; 4, outer membrane of serosa; 5, inner
membrane of serosa (only theoretical; it no longer seems to be developed); 6, primeval
mesoderm cell; 7, opening in the serosa (theoretical; according to Hatschek coalescence
here); 8, proctodaeum; 9, metatroch; 10, mesenteron. B, diagram of a 45 h-old embryo
of the lecithotrophic species *Golfingia elongata*. (After Åkesson, 1961a—modified.) 4,
cells of prototroch; other designations as in A.

impossible, the mouth opening being covered by the egg membrane as
well as by the serosa.

   This phase is, however, not of long duration; on the third day from the
start of the development the egg membrane and serosa are shed, and the
liberated *pelagosphaera* begins its planktotrophic life. The change is so

great that it might well be called a first metamorphosis. The real metamorphosis, i.e. the one that corresponds to that of other larval forms,
takes place at the transition to the benthic adult phase (see Jägersten,
1963). This metamorphosis is rather slow; there is no need for a rapid
process as an ample quantity of nutriment is stored in the stomach epithelium.

Gerould (1904, 1907) was the first to propose the interpretation
adopted here, that the outer membrane of the serosa is identical with the
prototroch cells of other *trochophora*-larvae. There seems to be no
valid objection to this proposal. It should, however, be pointed out
that the head of the planktotrophic *pelagosphaera* is provided with
a transverse band of cilia which on account of its position must be
interpreted as the prototroch of this stage (Fig. 39B). Here we are
probably faced with partial regeneration following the shedding of the
serosa.

In *Sipunculus nudus* the planktotrophic phase of life is said to last about
one month. During this time the larva attains a length of rather more than
1 mm. Several other forms reach roughly this size, though also some considerably larger larvae are known. The largest form that has been
described so far was found to measure 10 mm despite the fact that the
anterior end was retracted (Åkesson, 1961b). But even this record is sur-
passed by a still undescribed form which has recently been encountered in
the Indian Ocean. It was found to measure about 15 mm in the contracted
condition. Such giant larvae are probably much older than those of a
normal size. Unfortunately nothing is known about their metamorphosis. Possibly they are aberrant specimens which for some reason (perhaps disturbance of the internal secretion) have lost the ability to develop
into an adult and for this reason have continued to grow beyond normal
size.

As far as can be judged from Hatschek's investigation the *pelagosphaera*
type, when liberated from the egg membrane and serosa, already displays
the organization of the body which it then retains up to metamorphosis
into the adult. Apart from growth the only noticeable change seems to be
the storage of a considerable food reserve, mainly in the form of fat, in the
epithelium of the stomach.

It is remarkable that, in my experience, the giant larvae just mentioned
exhibit an exception in this respect. In them hardly any storage of food
can be traced, and this might indicate that such larvae do not form part of
a normal life cycle.

The *pelagosphaera*-larva exhibits characters of at least three different kinds. These are:

(1) *purely larval characters*, i.e. adaptations to the pelagic zone,
(2) *recent adult characters*, the formation of which has been shifted to the larval phase, and finally
(3) characters which cannot be classified under either of the first two categories and which must be interpreted as *ancient adult characters*.

Here, as in other larvae of the *trochophora* type, purely larval (pelagic) characters are primarily represented by the two trochi of the mouth zone. Of these the metatroch is particularly well developed after the first metamorphosis and in fact consitutes the real swimming organ. In the metamorphosis to the adult both are completely lost.

Recent adult characters include, among others, the U-shape of the alimentary tube, the position of the anus far forward on the dorsal side, and the division of the body into introvert and abdomen (see Fig. 39A and C) and the consequent ability of invaginating the former into the latter. These features are taken over by the benthic adult without any essential changes.

This is the case also with several other features such as the nature of the body wall, the spacious body cavity, the structure of the metanephridia, etc. These recent adult characters might be older than the first-mentioned, perhaps acquired already in the creeping ancestral form.

It is presumably the occurrence of recent adult characters in the *pelagosphaera*-larva that has prompted Damas (1962) to the declaration that the ontogeny of the sipunculids hardly goes beyond the *trochophora* stage, and that for this reason they ought to be considered as *trochophora*-larvae adapted to the benthic mode of life. This should explain the isolated position of the group. Apart from the elimination of the ancient adult characters the metamorphosis of the *pelagosphaera*-larva admittedly does not imply any more radical changes, but this is not a sufficient basis for the concept of Damas. The recent adult characters of the *pelagosphaera*-larva are by no means peculiar to the *trochophora*. There is no more reason to consider the sipunculids as derived from neotenic *trochophora*-larvae than e.g. the molluscs.

The most interesting kind of characters are the ancient adult ones. Among them is, first, the shape of the anterior end as a distinct head with a ventrally situated mouth. This presents a radical difference from the anterior end of the recent adult, with its ring of tentacles around the terminally situated mouth.

Other ancient adult characters are the ciliated lip or lobe, which is backwardly directed while in function (Fig. 39:1), the gland opening on this lobe and probably also the pharyngeal bulb (Hatschek's "*Schlund-kopf*", Fig. 39A). Like the purely larval characters the ancient adult characters are of course lost at metamorphosis.

By means of the ciliation upon the ventral surface of the head and the lobe the larvae are able to creep about upon a solid substrate. This behaviour is often observed in animals kept in dishes (see Fig. 39C). In the pelagic zone a creeping form of locomotion is of course out of the question. Unfortunately we still do not know where metamorphosis takes place. If it happens after the larva has come into contact with the biotope of the adult, it seems probable that the larva first creeps about upon the bottom for some time before it settles down permanently (cf. e.g. *cyphonautes*, p. 44). Whatever is the case in this respect I consider it likely that the creeping lobe and its ciliation as well as that on the ventral side of the head are reminiscent of a time when even the adult exhibited the same organization. They are evidently not characters which have been acquired by adaptation to the pelagic mode of life (see below).

This organization for locomotion also permits a special method of feeding. Unlike the adult the larva grasps and swallows individual particles— small living animals, such as other larvae, copepods, etc. The *pelagosphaera* might thus be called a predator. Since the mode of feeding is conditioned by the same adult characters which are also the basis for the creeping locomotion (see Jägersten, 1963), it is possible that this is a relic from the adult of the free-living ancestral form. In a *trochophora* the original feeding mechanism is quite different (see pp. 121 and 159).

The way in which the larva creeps about by means of a ciliated portion of the ventral side is in principle the same as that already known for certain other *trochophora*-larvae, e.g. in *Protodrilus* (Jägersten, 1940a, 1952). The main difference lies in the fact that the *pelagosphaera*-larvae have only a relatively short ciliated part immediately behind the mouth, while the majority of the other larvae have a narrow longitudinal ciliated band which extends over the whole distance between mouth and anus or the greater part of it (see p. 244). In the *pelagosphaera* the ciliated part is separated off as a special creeping lobe. (In this connection cf. also the larvae of the entoprocts, p. 105.)

In other *trochophora*-larvae the metatroch has a small discontinuity on the ventral side, and the longitudinal ciliated band runs through this gap. In quite a number, perhaps the majority of the *pelagosphaera*-larvae the metatroch is admittedly complete (Fig. 39A), but it is interesting to find that some species exhibit a ventral discontinuity, even if it is not very prominent (Jägersten, 1963).

The position of the metatroch far behind the prototroch and mouth (Fig. 39A) is a characteristic feature of the planktotrophic *pelagosphaera*-larvae by which they are distinguished from other *trochophora*-larvae. The position is obviously secondary. There is no doubt that the metatroch was originally situated so far anteriorly that the creeping lobe extended through the discontinuity mentioned.

What has been said so far applies exclusively to the planktotrophic *pelagosphaera*-larvae. In the larvae which belong to the lecithotrophic type the situation is considerably different. I shall now consider these (Fig. 41).

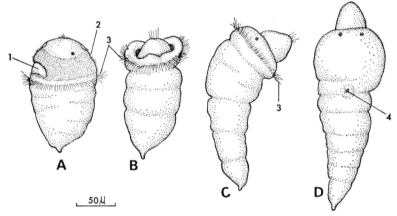

Fig. 41. Four stages of different age in the development of *Golfingia elongata*. A, 3 days; B, 4 days; C, 6 days; and D, 10 days after the beginning of development. 1, stomodaeum; 2, prototroch; 3, metatroch; 4, anus. Between A and B the disintegration of the cells of the prototroch takes place (cf. shedding of the serosa in *Sipunculus nudus*). (After Åkesson, 1961a.)

Here the pelagic phase is short, lasting only a few days, and during this time the larvae remain near the bottom. Of *Phascolion strombi*, Åkesson (1958) says for instance: "We can speak of a bottom-pelagic stage." In this respect the change has progressed farthest in the small *Golfingia minuta*, where the whole life cycle is benthic. The quantity of yolk in the egg and the length of the embryonic development have on the contrary increased. In *Golfingia minuta* the lecithotrophic phase lasts two months (cf. barely three days in *Sipunculus nudus*).

In connection with the increased quantity of yolk gastrulation has changed to epiboly and the lumen of the digestive tract has been lost in all species of this type examined so far, with the exception of *Golfingia elongata*, which according to Åkesson (1961a) has a "narrow, but definite archenteron".

Gerould (1904, 1907) is of the opinion that the development in the species *Golfingia vulgare* and *G. gouldi*, which is characterized by a short pelagic phase, relatively large quantity of yolk in the egg with consequent epiboly, and digestive tract without lumen, is the most primitive condition. Åkesson (1958, 1961a) adopts this view although his own new evidence argues against it. At the time when the larvae of the species mentioned are in "the stages within the spherical, unstretched egg-membrane" (1958, p. 64) they should agree best with the *trochophora* of the polychaets. This is correct so long as the comparison is restricted to insignificant external features or is made with larvae of polychaets which also have undergone alteration regarding their supply of food. But there is no point in a comparison of this kind.

In this comparison both Gerould and Åkesson have been led astray by the strong development of the prototroch and its transformation into a "serosa" in *Sipunculus nudus*. These authors are evidently of the opinion that this particular feature could not be primitive in a *trochophora*, and in this respect they are certainly right. But it is an error to make this feature the basis of the assumption that a short pelagic life, lecithotrophy, a large quantity of yolk, epiboly, and the absence of a lumen of the larval digestive tract should be characters which are original in the sipunculids. With regard to the occurrence of a lumen Åkesson realizes this error in his examination of *Golfingia elongata*, when he says quite correctly (1961a, p. 526): "However, the formation of an archenteron in *G. elongata* seems to be a real primitive feature." In my opinion the connection between large quantity of yolk and lack of lumen in the digestive tract of the larva or the embryo cannot be disregarded. In the sipunculids as well as in other groups of animals these two features are secondary, and they are also connected with lecithotrophy during some part of ontogeny—in the species which have been mentioned here, throughout the entire pelagic larval phase. And since in lecithotrophic larvae a long pelagic life is not necessary, this phase has been shortened, in the evolutionary line leading to *G. minuta* to such an extent that it has eventually resulted in a holobenthic mode of life. Åkesson is aware of the fact that in this respect this species is secondarily altered, but he has not come to the important and only logical conclusion of considering all sipunculid larvae exhibiting lecithotrophy and associated features as altered in comparison with the planktotrophic larvae.

A comparison between the two larval types of the sipunculids shows the organization of the lecithotrophic type to be simpler in several respects. The prototroch is admittedly broad and pronounced (Fig. 40B), but is not transformed into a serosa as in *Sipunculus nudus*. It has already been pointed out that the occurrence of the serosa cannot be considered as an original feature of the *trochophora*-larva as such, and the question arises whether it is original even in the sipunculids. If this should be the case, it would imply that at one time the serosa had existed in the larvae of all species, but had become reduced later on (in one or more instances) in the course of evolution to the now lecithotrophic forms. It is, however, possible that the occurrence of the serosa is restricted to the branch of the phylogenetic tree to which *Sipunculus nudus* and its closest relatives belong. This seems to be opposed by the observation that during metamorphosis of the lecithotrophic forms the cells of the prototroch perish (see below). Thus there is here a certain similarity with the fate of the serosa. Because of our limited knowledge, especially of the early ontogeny of the planktotrophic larvae, no final answer about the occurrence and phylogeny of the serosa can be given at the present time.

In the lecithotrophic larvae there exists no trace of the distinct head which is so characteristic of planktotrophic larvae. The same comment applies to the creeping lobe and its glandular organ and also to the pharyngeal bulb. It is, however, probable that at an earlier date these ancient adult characters were also found in the lecithotrophic larvae, *inter alia* for the reason that the characters in question must have been transferred to the pelagic larva at a time when the adult was free-living, i.e. before they disappeared in this stage as a result of the sedentary mode of life adopted at a later date. Once they had been eliminated in the adult they could not be shifted to the larva (cf. p. 222). Since it is not probable that the separation of the evolutionary lines leading to the lecithotrophic and planktotrophic forms respectively, had taken place during the non-sedentary phase of evolution, it seems likely that the lack of ancient adult characters in the lecithotrophic forms is secondary.

Also the occurrence of a strong metatroch and good swimming ability in *Golfingia elongata* and *G. vulgare* indicates that the lecithotrophic larvae are derived from forms which are more like the planktotrophic ones. In this respect these *Golfingia* species differ from *G. gouldi* and *Phascolion strombi* where the metatroch is considerably reduced, and still more from *G. minuta* in which the whole ciliation is changed.

This reduction of the rings of cilia together with the absence of the ancient adult features characteristic of the *pelagosphaera* type (distinct head, creeping lobe, etc.) in the lecithotrophic forms means that the ontogeny of these latter must be considered as being on the edge of direct development. One might even go so far as to say that in the case of *Golfingia minuta* the term direct development is the most appropriate designation.

Åkesson (1961a) reports that in *Golfingia elongata* metamorphosis takes place when the larva is roughly 48 h old. Åkesson evidently applies the term metamorphosis primarily to a certain lengthening of the body (Fig. 41). The pelagic mode of life is not terminated by this change. For some days afterwards the larvae swim about by means of the metatroch. In my opinion this later phase of pelagic life might be the shortened counterpart of the long planktotrophic period in the *pelagosphaera*-larvae, which also move about during this period by mean of the metatroch. It is thus possible to say that in the lecithotrophic forms the two metamorphoses, which in the *pelagosphaera*-larvae are widely separated in time, have come nearer each other. At the same time the general paucity of characters has resulted in a simplification of these metamorphoses, implying practically direct development. It is a remarkable fact that the body is enclosed by the egg membrane for a considerable time—longest in *Golfingia minuta* (this is connected with the embryonic retardation ''resulting'' from the large quantity of yolk).

While in *Sipunculus nudus* (and presumably also in other planktotrophic forms) the cells of the prototroch have been transformed into a serosa, they have undergone a different change in the lecithotrophic forms. In these they contain, strangely enough, a large quantity of yolk grains which, when the cells degenerate, find their way into the body cavity where they are consumed. A situation of this kind in which the cells of the prototroch constitute a nutriment reserve can be no more primitive than their transformation into a serosa in *Sipunculus*, but must have arisen in con-

nection with an increase in the quantity of yolk and evolution towards
lecithotrophy. As has been pointed out, we are still unable to decide
whether or not the ancestors of the lecithotrophic forms had the cells of
the prototroch transformed into a serosa. Whatever the case may be,

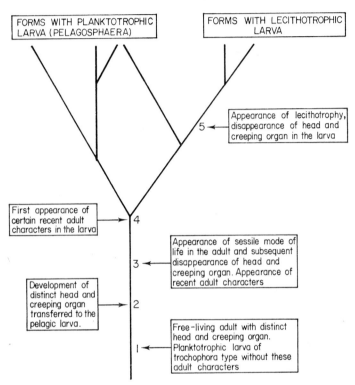

Fig. 42. Diagram indicating the approximate position of important features and changes
in the life cycle during the course of evolution leading to the recent *Sipunculida*. It may be
imagined that the changes at points 4 and 5 have occurred independently in several
evolutionary lines. There is, however, no direct evidence. (The diagram is not intended
as a phylogenetic tree in the sense that it represents the phylogeny of the different systematic units within the group.)

these cells are now changed, although in different ways, in all sipunculids
that have so far been closely examined. The original situation was a quite
simple prototroch.

It should finally be stressed that the changes in the life cycle in the
course of evolution up to the present are on the whole analogous with
those in other sedentary groups. As a result of adult pressure the initial

development of adult characters, ancient as well as recent, have been shifted to the pelagic larva. (For a more detailed discussion of the principles concerning changes of this kind, see p. 5 *et seq.*)

In Fig. 42, the approximate stages at which the transfer of some important adult characters to the pelagic larva took place are indicated. This diagram represents a summary of certain points discussed in the foregoing text. By way of comment it might be added here that the planktotrophic larva of the ancestor in point 1 was probably feeding by means of a ciliated adoral groove of the same kind as in the present primitive *trochophora-*larvae (see pp. 108 and 159), and that among the recent adult characters in point 4 the position of the anus far forward on the dorsal side is an important external feature.

Myzostomida

All myzostomids are commensals or parasites on or in benthic animals (echinoderms, almost exclusively crinoids).

It is known that a larva of the *trochophora* type occurs within the genus *Myzostomum* (Jägersten, 1939a). This has admittedly never been caught in the plankton, but its ciliation and the long provisional chaetae (Fig. 43) make it evident that it must lead a pelagic life at least for a short time. This larva develops from small eggs, poor in yolk, which are discharged freely into the water. Although in this respect the primitive condition has been retained the eggs are nevertheless fertilized when discharged. (On the unique mechanism for copulation in these animals, see Jägersten, 1939b.)

The presence in the mature larva of both mouth and anus and a highly differentiated digestive tract with continuous lumen is related to the small quantity of yolk in the egg. Unfortunately nothing is known about the feeding mechanism or about the developmental phase which immediately follows the stage pictured in Fig. 43, but unless the larva changes almost immediately to parasitic life this must be a planktotrophic form. Planktotrophy must in every case have occurred within the genus *Myzostomum* in the not too distant past, measured in terms of phylogeny.

In certain endoparasitic genera, on the other hand, ontogeny seems to be more altered. Here the earliest phase is unfortunately entirely unknown. The youngest observed stage is already essentially like the adult (see Jägersten, 1940c, fig. 12), and for this reason, bearing in mind that there is a certain mass of yolk in the egg, it appears that development might be direct. No conclusive statements are, however, possible at the present time, and for this reason further investigations into the ontogeny of the various families would be highly desirable. From what we know

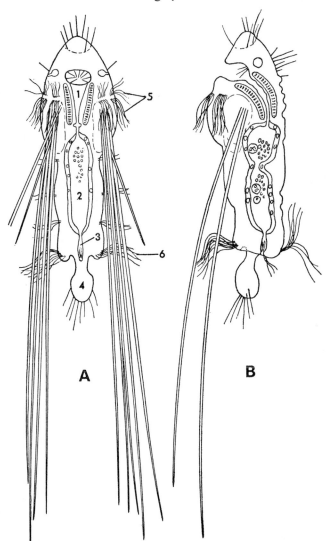

FIG. 43. Larva of *Myzostomum parasiticum*. Oldest stage obtained in culture. In B the greater part of the floating setae has been lost. 1, stomodaeum, surrounded by muscular muff; 2, mesenteron (stomach); 3, hind gut; 4, tail appendage; 5, rudiment of metatroch; 6, telotroch. (After Jägersten, 1939a.)

about the pelagic larva it exhibits no traits which could be interpreted as ancient adult characters or their remains.

Annelida and Echiurida

There is considerable variation in the life cycles of these groups. In Polychaeta (including Archiannelida) and Echiurida the usual rule of pelagic primary larva and benthic adult applies, even if in certain cases a holopelagic mode of life has evolved, while in others the larva has been replaced by benthic direct development. The latter change has taken place without exception in the groups Oligochaeta and Hirudinea which have emigrated into the fresh water or even on to dry land. Here direct development implies profoundly altered features, in certain cases in connection with a special nutrient supply in the form of solutions of albumen secreted in the cocoons by the mother animal. Because these non-marine groups are not particularly interesting in the present discussion they will not be considered in any further detail (see, however, pp. 172 and 239).

The marine groups are characterized by the well-known *trochophora*-larva which has, however, in several cases been replaced by direct development here, too (see below). The rings of cilia are the only conspicuous pelagic characters of this larva. For this reason metamorphosis is as a rule not very far-reaching. Exceptions occur, however, such as the well-known endolarva of *Polygordius* and the *mitraria*-larva of the ammocharids (Wilson, 1932). In connection with a lengthy pelagic life these forms have acquired such peculiar morphology that large parts of the body have to be dropped at metamorphosis. The changed morphology in *Polygordius* consists of a strongly "inflated" body (the blastocoel is particularly enlarged) and in the *mitraria*-larva a hypertrophied oral zone extends in bights.

Different opinions have been expressed about the metamorphosis of the "endolarva" of *Polygordius*. According to Woltereck (1902, 1905, 1925) practically everything between the apical plate and the segmented trunk is shed. Söderström (1924a, b; 1925a) categorically refutes such a "catastrophic" process and asserts instead that there is a gradual change. I cannot enter more closely into the controversy between Woltereck and Söderström; I only want to point out that the former was undoubtedly right about the main question (see also Wilson, 1932). I have had several opportunities to observe not only that the oral zone with its cilia rings is shed, but also that the dropped parts are ingested by the young worm, which in this way is secured an ample first meal (cf. p. 217). This observation has been made on *Polygordius appendiculatus*, where after metamorphosis the pigmented cells of the prototroch can easily be seen in the enteron.

Contrary to the attitude adopted in handbooks and textbooks Åkesson (1962) asserts that the larva of *Polygordius* cannot be referred to as a *trochophora* representative of the annelids. In part this is quite correct. With its spacious blastocoel and radical metamorphosis it is neither representative nor primitive. The polychaets usually exhibit a more narrow blastocoel and, in relation to this, a smaller body size and more gradual metamorphosis (without shedding anything other than the provisional chaetae and certain cilia). The larva of *Polygordius* is, however, original in its feeding apparatus (see below).

Åkesson also considers the larva of *Eupomatus* (= *Hydroides*) as hardly typical, following the view expressed previously by Shearer (1911). I feel personally that the young *trochophora* of this genus is fairly typical of the original larva. Admittedly it does have a relatively wide blastocoel, but the body is not transformed to the extent that any larval parts have to be dropped at metamorphosis. In general the spaciousness of the blastocoel is characteristically associated with the quantity of yolk in the egg. In forms with a rich supply of yolk (and lecithotrophy following from it) the blastocoel is as a rule more or less narrowed by the enlarged entoderm. In planktotrophic larvae the wall of the mesenteron is fairly thin, at least initially, and particularly if the eggs are very small.

The *trochophora*-larvae of this type possess a special feeding apparatus which was already known to Hatschek (1885). In *Polygordius* as well as in *Hydroides* and certain others, e.g. *Pomatoceros* (Fig. 44), the zone between the strong prototroch and the feebler, postoral metatroch bears a very fine ciliation which ventrally joins the cilia of the mouth and the oesophagus. Hatschek calls this ciliated zone "adorale Wimperzone", while Segrove (1941) speaks about "feeding cilia". Both terms are very appropriate. I feel that Segrove has arrived at the correct conclusion about the feeding mechanism, when he says that the cilia of the prototroch push the particles down to these feeding cilia which then transport them to the mouth. It can be added that only small particles can be ingested in this way; their passage through the oesophagus does not as a rule necessitate its widening.

I have paid particular attention to this mode of feeding and have found exactly the same mechanism in widely separated groups of animals (see pp. 108 and 121). For this reason and because of its wide distribution within certain of these groups it appears to me inevitable that this mechanism should be considered as original not only for the Annelida but

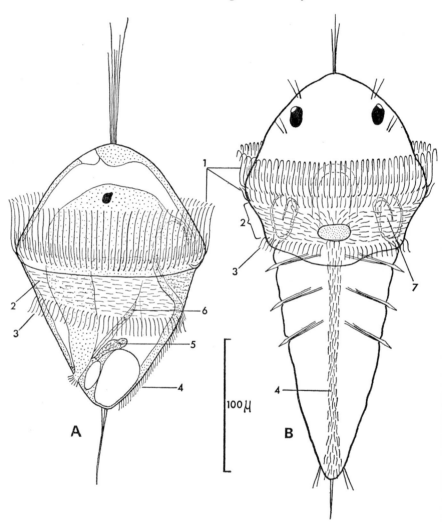

Fig. 44. *Pomatoceros triqueter*. A, *trochophora* (about 6 days); B, *metatrochophora*, fully developed (about 3 weeks). 1, prototroch; 2, adoral ciliated zone; 3, metatroch; 4, ventral band of cilia ("neurotroch"); 5, mesoderm strand; 6, protonephridium; 7, "collar". (After Segrove, 1941, partly simplified.)

also for the Entoprocta and Mollusca (and thus in all probability for all *trochophora* groups).

Since planktotrophic larvae of the type described always develop from small eggs which are poor in yolk and which as a rule are discharged free, i.e. without cocoons or other envelopes, I consider that this *combination of features* is an indication that these features are indeed primitive (see also p. 240).

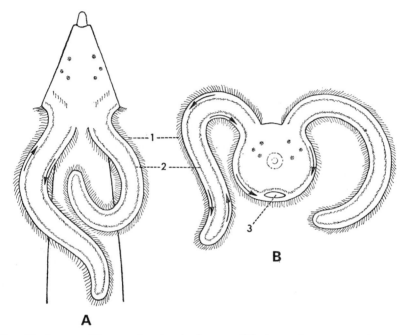

Fig. 45. *Rostraria*. Diagrams showing the features of the tentacular apparatus. A, anterior end of the larva in dorsal view; B, front view, the tentacles projected into the plane of the paper. 1, prototroch; 2, metatroch (actually hidden); 3, mouth (actually hidden). The arrows indicate the direction of the transport of food particles (cf. Fig. 26A and B).

In the polychaets I have been able to make direct observations of the ingestion of food via the adoral zone of cilia *inter alia* in a form of the peculiar "rostraria" type (Florida current at Bimini, end of February, 1965). In this larva the long tentacles assist in the process. As far as I can see these tentacles are nothing more than greatly extended dorso-lateral lobes of the body zone which bears the adoral ciliation and the proto- and metatroch. Thus these tentacles differ from the velar lobes in a molluscan larva only by their narrow shape (cf. Fig. 45 with Fig. 33A). By the action of the adoral cilia the food particles are unfailingly carried into the mouth, regardless of where they are caught on the tentacles. If the capture takes place somewhere on the dorsal side the particles are first transported to the tip and then along the ventral side to the mouth (see the arrows in Fig. 45).

A more detailed account of the *rostraria*-larvae, which are still unknown in many respects, falls outside the scope of this book. I want, however, to stress that my observations clearly show the shortcomings of Häcker's (1898a) speculations about the tentacles. He thinks that they have nothing to do with feeding, but are rather some kind of "Stossfühler" . . . "welche abwechselnd eingezogen und ausgestreckt werden können und damit dem Tiere eine aktive, stossweise Bewegung verleihen". These unfounded ideas of Häcker have been accepted in various textbooks. The larvae which I studied did not in a single case exhibit more than very feeble and slow movements of the tentacles; in general these remained perfectly motionless. Swimming was found to be performed by means of a strong perianal ring of cilia (telotroch). (In a future paper I hope to give a more detailed account of this *rostraria*.)

The tentacles of the *rostraria*-larvae and the velum of the molluscs have of course originated independently in their respective groups. All the evidence seems to indicate that the occurrence of velar structures is not the original situation either for annelids or for molluscs (cf. p. 140 *et seq.*).

The mechanism for feeding by means of adoral cilia must be very ancient, as has already been pointed out. To what extent it is retained in the larvae of the annelids has not yet been determined in any detail.

It is, however, evident that it has been lost in the majority of forms. This applies not only to lecithotrophic larvae and larvae which swallow larger particles (cf. the larvae of the palaeonemertines and the sipunculids, pp. 90 and 150) but also to larvae which still feed on smaller particles. An example of the latter is *Protodrilus* (Jägersten, 1952). There the adoral ciliary apparatus has been replaced by a complicated mechanism by which the oesophagus is extroverted in the shape of a catching funnel when food is ingested (Fig. 46A). (In the larva of *Pectinaria* the margin of the mouth has become protracted into a structure which exhibits a superficial similarity.) It is impossible to say why such a change has taken place unless one is prepared to offer the usual "explanation", that the new condition is more advantageous than the original one. It is more understandable, and, so to say, as it should be, that the adoral ciliary

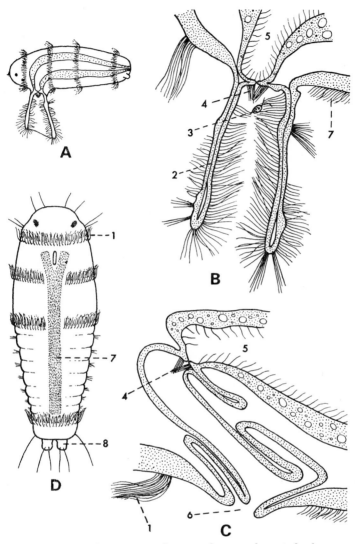

FIG. 46. *Protodrilus rubropharyngeus*. A, larva in the growth period, the stomodaeum extroverted as catching funnel. B, catching funnel in greater detail. By the long cilia of the funnel the food particles are whisked towards the opening between stomodaeum and mesenteron, where they are swallowed by the latter. At rest the opening is closed and covered by sensory hairs. C, stomodaeum at rest, folded in between mouth and mesenteron (cilia of the stomodaeum are not shown). D, fully developed larva in ventral view. 1, prototroch; 2, the double wall of the funnel; 3, food particle; 4, group of sensory hairs around the opening to the mesenteron; 5, lumen of the mesenteron; 6, mouth; 7, ventral band of cilia ("neurotroch"); 8, adhesive lobes. (A–C after Jägersten, 1952; D original.)

apparatus has been lost in lecithotrophic larvae and in larvae which have started to ingest larger particles.

The larvae of the echiurids exhibit very interesting features in this respect. Although some of these larvae swallow large particles they may

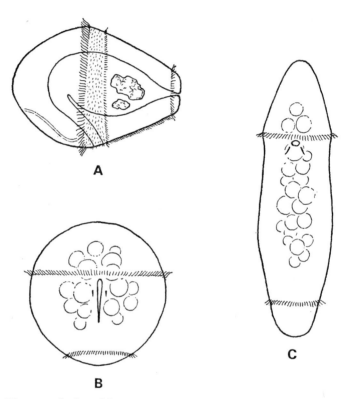

A

B

C

FIG. 47. Diagrams of echiurid larvae. A, young specimen of planktotrophic species with metatroch, adoral cilia, and longitudinal ventral band of cilia ("neurotroch") retained. Stomach epithelium still without stored nutriment. Ingested food particles in lumen. B and C, stages of a lecithotrophic species without any trace of the ciliation mentioned in A and without continuous gut lumen. In the mesenteron cells plenty of fat, derived from the yolk of the egg. Near the mouth the two hooked setae are visible. (About this species see footnote on next page.)

possess both metatroch and adoral ciliated zone. This observation was made by Hatschek (1881) on an *Echiurus* sp. from the Mediterranean. I have been able to verify the point in relatively young larvae of a species from the Florida current (Fig. 47A). These were found to be able to ingest larvae of sipunculids almost as large as themselves. In spite of lengthy observations of several specimens I could not find any indication in this species of the ability to transport particles by the adoral ciliated zone. It is in any case unlikely that objects as large as the larvae of sipunculids could be caught and transported to the mouth by the cilia in the adoral zone. Evidently this adoral zone has no function in feeding. For this reason it is rather surprising that it has not become reduced; perhaps it functions in the very young stages. Under any circumstances it is worthy of comment that an adoral apparatus is also found in the echiurids, since this gives support to the proposal that such an apparatus represents the original condition in the annelids as well as in certain other groups (see pp. 108 and 121).

It seems, however, that the echiurids do not always conform to the conditions described, since I have also encountered a larva with no trace of either metatroch or adoral cilia. The absence of these features may be connected with the occurrence of lecithotrophy in this larva.*

---

* I should like to add some further comments about the lecithotrophic echiurid larva in question, the species of which is unfortunately unknown. It was found in great numbers towards the end of February 1964 in surface water near the shore at Mandapam in southern India. Most freshly-caught specimens were in the stage represented in Fig. 47B, and characterized by the spherical shape of the body, a simple prototroch slightly anterior to the middle, and a telotroch at the posterior end. In the middle of the ventral side is a longitudinal fissure which is widest at the front. This is obviously the blastopore which is being closed progressively from back to front. All over the surface except below the rings of cilia the epidermis bears small green spots. The larvae swam or floated with the anterior end upwards immediately below the surface of the water, their buoyancy being maintained by a small number of remarkably large fat globules in the entoderm.

After one day the larvae had already lengthened considerably and assumed a shape more typical of the larvae of echiurids (Fig. 47C). At this stage some of them began to creep about on the bottom of the dishes, alternating with swimming. The mouth was represented by a very small rounded depression close to the prototroch. The oesophagus still had no lumen. The hooked setae characteristic of the group were developed a short distance behind the mouth.

I was only able to continue the study of these larvae for one additional day. On this day an increasing number of specimens started to creep. The pelagic period evidently lasts only a few days. In one specimen in which development had evidently progressed somewhat further (here the telotroch was missing) it could be observed that locomotion was performed by means of fine cilia on the ventral side of the prostomium, possibly assisted by contractions and extensions of the body.

This reasoning leads to the conclusion that all fully grown larvae of annelids and echiurids which are without adoral cilia are secondary in this respect. Thus I am forced to refute Segrove's (1938, 1941) views that *planula*-like, "atrochous" larvae are more primitive than those with an adoral apparatus. Larvae with more uniform ciliation have been observed, e.g. among Eunicidae, and in all known cases they are lecithotrophic. Because, in a number of species of polychaets the embryos leave the egg with uniform ciliation all over the body which is then differentiated into the usual trochi and adoral cilia (e.g. *Mercierella enigmatica*, according to Rullier, 1955) it is probable that those larvae that are atrochous even in older stages have lost this differentiation, which in lecithotrophic forms seems quite unnecessary. In other words, the atrochous lecithotrophic larvae seem to have retained an embryonic ciliary organization. This is in accordance with the statements of Anderson (1966), that no apical tuft of cilia, no larval musculature, and no protonephridia are developed in lecithotrophic larvae.

In maintaining that the more or less *planula* like, atrochous larvae of polychaets are altered, *inter alia* by the loss of the adoral apparatus, I do not mean to exclude the possibility that a uniform ciliation covering the entire surface of the body is a very primitive feature of the Coelomata—a feature which is retained, for example, in certain polychaets (see also pp. 190 and 209). Could this similarity with the *planula*-larva of the cnidarians be phylogenetically conditioned? I do not consider it impossible.

As stated above, metamorphosis of the *trochophora* follows a fairly even course in the majority of the annelids, even though the transition to the benthic phase is accompanied by a number of changes. As far as external appearance is concerned this metamorphosis consists first of all of an elongation of the body and the loss of the larval ciliation and of provisional chaetae, if they are present. Some other less decisive changes take place after the transition to the benthic zone.

In the majority of the spionids and some relatives of this family (*Poecilochaetus*, *Disoma*, *Magelona*), and in representatives of several other

---

It was striking that the larvae represented two size groups with roughly the same number of specimens in each. In the larger specimens the spherical stage had a diameter of about 0·6 mm, in the smaller only half that size. No other difference could be established with certainty. Could this be a case of sexual dimorphism? A more detailed investigation would be welcome.

families we find an extended pelagic existence combined with considerable growth of the body, but in these forms, unlike *Polygordius*, the adaptation of the larva to the pelagic zone has meant only a slight divergence from the adult. Instead the organization of the growing larva gradually changes to become more like that of the adult. Here the effect of "adult pressure" is such that one could almost speak of direct development. This term would nevertheless not be applicable, since also in this type a slight metamorphosis takes place at the transition to the benthic zone. This is also the case in those forms where the progression towards the organization of the adult has gone so far that the development of sexual cells has started during the pelagic phase. Evolution in this direction has been possible because the combination of larval and adult organization which occurs in the forms in question is suitable for a pelagic mode of life. This is almost a truism, but must nevertheless be taken into account, since it may be a step towards the holopelagic mode of life that has been adopted by quite a number of polychaets (*Tomopteris*, *Alciopa*, etc.). The cases of lengthened pelagic existence and development of sexual cells established in larvae of certain spionids (Hannerz, 1956; Berkeley and Berkeley, 1963) could perhaps be interpreted as an "attempt" at evolution in this direction.

However, I want to point out that as far as I can see there is no direct indication that the holopelagic polychaets have evolved via a lengthened pelagic larval life. It appears more likely that their ancestral forms in the *adult* stage began to perform swimming movements in alternation with creeping locomotion and finally left the bottom altogether. (Cf. the holopelagic gastropods, p. 130.) It might be asked whether or not the change to a holopelagic life cycle could have any connection with the brief pelagic life found in certain benthic polychaets during propagation (epitokism). But this is a problem which cannot be dealt with here.

According to Friedrich (1949), in the majority of the holopelagic polychaets the *trochophora* has been replaced by direct development. If this is the case it would only indicate that the same rule applies to polychaets as to other groups (see p. 236).

Like other primary larvae the annelid *trochophora* exhibits adult (benthic) characters in addition to the purely larval (pelagic) ones. In general they are only in their first phase of development, e.g. parapodia, antennae, palpi, and other appendages of different kinds. This applies on the whole also to the chaetae, but among them the provisional ones, the "larval" chaetae, form an exception in that they are lost at metamorphosis. Nevertheless I consider that these provisional chaetae are in reality an adult character. We have here an example of the by no means unusual phenomenon that an adult character, transferred to the primary larva, has acquired a special use (floating chaetae!) and in connection with it has become modified and must for this reason be replaced at metamorphosis (cf. p. 223).

In the early stages of development the *trochophora* of the annelids is completely unsegmented, and already this seems to suggest that the segmentation is an adult character, originally restricted to the benthic phase of life. It might be imagined that in the first annelids the acquisition of a segmented body took place during metamorphosis or perhaps after the transition to the benthic zone. In my opinion even the first segmentation appearing in the pelagic larva ("primary segmentation") can be explained as a result of "adult pressure". In several forms further steps in the same direction have been taken, i.e. additional segments have been acquired by the larva, and in relation to this the pelagic phase has been extended, as is the case in *Magelona* and the majority of spionids (see also below).

The hypothesis that the occurrence of segmentation in the pelagic larva can be interpreted as a result of adult pressure of course offers no explanation of the frequently discussed dissimilarities between "primary" and "secondary" segments (Iwanoff, 1928). These dissimilarities are conditioned by other phylogenetic changes which probably also found their first expression in the benthic phase of the life cycle.

I consider it probable, however, that the shifting of the development of primary ("larval") segments to the pelagic phase took place in an evolutionary stage in which the adult annelid had no more than these few segments, i.e. before the appearance of the pygidial zone of growth. After this zone had arisen and the adult had acquired secondary segments, in certain cases (examples have already been given) part of these phylogenetically younger segments were also transferred to the pelagic larva.

The different number of primary segments which has been established in various forms and groups may possibly indicate that the transfer to the larval phase has taken place independently in different evolutionary lines. But the possibility cannot be excluded that in certain cases a change in number might have taken place in the larvae at later stages of evolution.

Segmentation of the larval body is a general feature. Other recent adult characters are as a rule restricted to systematic units of lower rank. Often it is a question of appendages and processes of different kinds as has already been mentioned. Here only a few examples shall be given.

The characteristic adhesive lobes on the pygidium of *Protodrilus* are developed at a very early date, in certain species even in the newly-hatched larva (Jägersten, 1952). This early development is quite unnecessary since the lobes are not put to use until after metamorphosis.

In forms within Aphroditidae, Phyllodocidae, Spionidae, Sabellidae and other families, tentacles and palpi of different kinds occur at the anterior end in the pelagic phase. It is possible that these structures, too, have no function in most larvae, but in certain cases where the pelagic phase is lengthened they are put into use. A marked example is the larva of *Magelona* in which the long palpi are efficient instruments for catching the prey. (In this case and similar ones it would be of value if a more detailed examination were carried out to determine whether, and to what extent, their morphology and function is altered at the transition to benthic life.)

The so-called collar of the sabellids and serpulids is another recent adult character which starts its development in the pelagic larva. In *Pomatoceros* this structure is developed in a pocket (Fig. 44B), evidently an adaptation to the pelagic mode of life. In *Spirorbis* the development of the collar is initiated already during embryogenesis. This is a result from the lengthening of the embryonic life at the expense of the pelagic period.

An instructive example of the transfer of the development of a benthic adult character to the pelagic phase is provided by the terebellids, e.g. *Lanice*, where the secretion of the tube begins long before the transition to life on the bottom.

In other cases, on the other hand, development of adult characters may start only at or after the metamorphosis proper, i.e. in the biotope of the adult. This applies *inter alia* to the tentacles and palpi (tentacle crown) in, e.g. *Protodrilus* (Jägersten, 1952), *Pomatoceros* (Segrove, 1941), and *Mercierella* (Rullier, 1955). In such cases "adult pressure" has thus had no effect. This is remarkable in view of the fact that in related forms, e.g. *Branchiomma* (Wilson, 1936), the same organs have already started to develop before the transition to benthic existence (Fig. 48). At present the cause of this difference cannot be given. The fact that in many cases the appendages in question are developed only after the pelagic phase is, however, of interest from the theoretical point of view, since it supplies an argument in favour of the thesis presented in this book, that the first

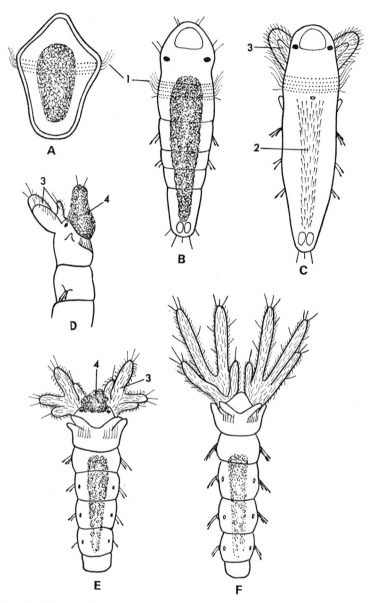

FIG. 48. Developmental stages of *Branchiomma vesiculosum*. A–C, swimming larvae (C ready for metamorphosis); D–F, young benthic stages extracted from their tubes. 1, prototroch; 2, ventral band of cilia ("neurotroch"); 3, crown of tentacles; 4, part of anterior end which disintegrates and is lost at metamorphosis. (After Wilson, 1936, somewhat simplified.)

phylogenetic appearance of the benthic adult characters must always take place in the benthic phase (see also p. 220 *et seq.*).

A more detailed study of the adultation of the palpi and other appendages and processes in the polychaets would be of interest. Already it seems possible, however, that there have been cases of parallel adultation, i.e. that one and the same organ has been shifted to the larva independently in different evolutionary lines (cf. also p. 56).

It appears that parallel adultation might explain an apparently paradoxical situation in the *Polychaeta*. Anatomical examinations of the nervous system by Orrhage (1966) have led to the conclusion that the long catching tentacles in *Spionidae* and closely related families are homologous not only with the palpi of the errantians but probably also with the crown of tentacles in *Sabellidae* and *Serpulidae*. This homologization, however, meets with the difficulty that in the spionids and their relatives the tentacles develop *behind* the prototroch, while in the errantians (all of them ?) the palpi and in the sedentarians mentioned the crown of tentacles are formed *in front* of this ring of cilia (see Fig. 48). This certainly poses a problem. If the homologization is correct, the most likely explanation seems to lie in the assumption that the transfer to the larva of the initial development of these palpal structures has taken place independently at least twice, firstly in forms where the palpi are situated comparatively far backwards (spionids, etc.) and secondly in those where they occupy a more anterior position. With regard to these structures adultation thus seems to have occurred regardless of the location of the purely pelagic character of the prototroch. The possibility of this interpretation means that a homology between these structures does not necessarily signify that in certain larvae the palpi have moved through the prototroch during the course of evolution.

A very important adult character is the longitudinal ventral band of cilia ("neurotroch") which has been established in the larvae of all major groups of annelids. In the polychaets this band occurs in the majority of species. In *Protodrilus*, where it is particularly well-developed and distinct, it extends from the mouth to the telotroch (Fig. 46D). In certain forms, e.g. *Pectinaria*, the band runs through a small discontinuity in the telotroch up to the vicinity of the anus.

In *Protodrilus* and a few other small archiannelids and polychaets the band also occurs in the adult and functions as an organ of locomotion (creeping organ). During the pelagic phase, on the other hand, it of course has no such function. In these forms the band must be characterized as a typically recent adult character.

In the numerous species in which this ciliated band is restricted to the larva it must, on the contrary, be interpreted as an ancient adult character, a relic from the time when the respective ancestral forms moved in the same way as *Protodrilus* does at present.

I want to leave as an open question the possibility that in the evolutionary lines leading to *Protodrilus* and others, the band may have disappeared in the adult, and that later on, in connection with a decrease in size of the adult, it may have been taken over from the larva. In such a case the adult might be termed neotenic concerning the band. I can, however, find no indication of a course of this kind. This possibility, in any case, does not affect the interpretation that the band is an ancient adult character in those cases where it occurs solely in the larva.

In forms where the adult is without the ciliated band this is lost at metamorphosis. It is interesting to find that in some larvae, e.g. that of *Pomatoceros* (Segrove, 1941) the band functions as a creeping organ for a short time before sedentary life is adopted (cf. the pear-shaped organ in the bryozoans, p. 44). In other larvae this function seems to have been lost.

The fact that remains of the band are still present, at least in some cases, in the embryos of the groups Oligochaeta and Hirudinea, which otherwise show strongly altered ontogeny, is proof of the tenacity with which this ancient adult character is retained.

It appears that the ventral band of cilia is the only commonly occurring ancient adult character of a positive nature in the annelids. It is, however, most interesting that the same band is also found in a number of other phyla. This is evidently a character which is much older than the annelids themselves (see p. 244).

It must be stressed that a pelagic larva occurs in most species of the marine annelids. According to Thorson (1946), using data from the region of Öresund, such a larva is found in 70 per cent of the polychaets for which the propagation has been studied. In many species, although perhaps not in the majority as maintained by Gravier (1923), the sexual products are discharged freely into the water. These observations indicate that development via a pelagic *trochophora* is original for the annelids. Another indication is that distinct remains of the trochi (these are purely adaptive characters to pelagic life) are also found in the larvae of holobenthic forms which usually show some kind of brooding (development in cocoons and other envelopes is included here). In the larva of the terebellid *Nicolea zostericola*, for instance, which leaves the cocoon at the *metatrochophora* stage both proto- and telotroch are retained in an almost unaltered condition in spite of the fact that the larva never leaves the bottom. *Scoloplos armiger** and *Protodrilus symbioticus* can be mentioned as additional examples, in spite of the fact that at least in the latter the development has been altered to such an extent that it might almost be called direct (Swedmark, 1954). Ciliation which without doubt represents the remains of the rings of cilia of the *trochophora* have been found on either side of the mouth even in the still more transformed embryos of certain oligochaets. It would be unreasonable to interpret these and similar features in other embryos and non-pelagic larvae as ''preparations'' for a future pelagic life. Here, as for example in certain gastropods, the incomplete rings of cilia show clearly that at one time the ancestral forms had pelagic larvae.

Within the Polychaeta (I am also including here the Archiannelida) we find practically all transitions from larvae with a long planktotrophic life developing from small eggs, that are freely discharged into the water, to direct development in the benthic zone. (This last remark applies, e.g. to *Dinophilus*, see Jägersten, 1944.) The transitional forms are the lecithotrophic larvae which swim about freely for a longer or shorter time. The changes have taken place independently in many different instances. A closer analysis is unavailable at present, but would be most interesting.

The state of instability within the life cycle can be seen, for example, in *Polymnia nebulosa*. According to Häcker (1898b) the larvae in most cases leave the gelatinous envelope of the egg mass at the *trochophora* stage and live pelagically for some days. But frequently they remain inside the envelope for such a long time that by the time they leave it they possess up to six segments carrying chaetae. In this case they can immediately start to build their tubes. This is evidently a species which is close to adopting a holobenthic mode of life. Rasmussen (1956) gives other examples of species which in one and the same locality can exhibit both

---

* *Scoloplos armiger* is mentioned here as a holobenthic form on the basis of a statement by Thorson (1946) concerning Öresund. According to Sveshnikov (1960), using data from the White Sea, the species has a pelagic larva. This indicates that the transition to holobenthic life is so recent that it has taken place in the existing species.

pelagic and non-pelagic development, viz. *Nereis pelagica, Pygospio elegans,* and *Capitella capitata.*

It has been mentioned above that forms with a holobenthic life cycle frequently still retain traces of pelagic characters (structures of the primary larva). *Manayunkia,* however, is an example where these have been completely lost (cf. p. 180).

With the possible exception of cases in which the mother animal provides the offspring with nutriment within the cocoon, the eggs of forms with lecithotrophic larvae, or of forms with direct development are in general considerably bigger and richer in yolk than the eggs of species with a planktotrophic *trochophora.* There is nothing to contradict the assumption that small eggs and planktotrophy represent the original condition, and that lecithotrophy has arisen secondarily in different instances in connection with an increased quantity of yolk. There are, on the contrary, several pieces of evidence pointing in this direction. For one thing, the great majority of eggs which are discharged to float freely without brooding—absence of brooding should be considered as the original situation—develop into planktotrophic larvae. (Intimate brooding is on the other hand most often connected with lecithotrophy.) Further: primitive features include a ciliated coeloblastula, ciliated gastrula, gastrulation by invagination, and feeding by means of an adoral ciliated zone. The fact that this mode of feeding is found also in Mollusca and Entoprocta is perhaps the strongest argument in favour of the primitive nature of planktotrophy in these three groups.

All the foregoing evidence is in accord with my view that a pelagic larval phase and benthic adult phase (pelago-benthic life cycle) is the original situation in the annelids. There is nothing to support the opposite possibility, i.e. that juveniles of holobenthic forms passed secondarily into the pelagic zone and were transformed there into larvae. The *trochophora*-larva is much older than the annelids. It existed even in the last common ancestor of all *trochophora* groups. The matter of the shape of this ancestor in the adult stage will not be discussed here. I shall only point out that the adult did not resemble the *trochophora.* This is solely a larval form and has never been anything else.

Crustacea

Within the great complex of arthropods the primary larva is retained only in the Crustacea, which are in many respects the most original group. In all other groups a change to direct development has taken place. This direct development has, however, in many cases been complicated by new divergences between the juvenile and the adult, leading to secondary larvae and secondary metamorphoses. This is particularly pronounced in the insects. In the terrestrial arthropods the situation is in some respects, though not as a principle, different from that in the groups which still inhabit their original environment—the ocean. For this reason I shall restrict myself to a discussion of the crustaceans.

Within all the great subgroups of the Crustacea the *nauplius*-larva is represented. Since this is so, and since the crustaceans in some fundamental features agree closely with the annelids, it is certain that this larva can be considered as the original type for the entire arthropod complex. In other words, it can be stated that the *nauplius* is derived from the *trochophora*. Ever since its appearance it has accordingly been pelagic and radically different from the benthic and originally worm-like adult. The *nauplius* is a "primary larva" just as much as the *trochophora*, even if it has undergone considerable alterations by adultation (see below).

The differences between the *nauplius* and its adult are roughly of the same order as those between the *trochophora* and the adult belonging to it. A different way to express this would be to say that the divergence between the two phases of the life cycle has hardly increased at all since the time when the crustaceans were separated from the annelids.

In the case of the *nauplius* is it perhaps more evident than for any other primary larva that it cannot recapitulate the organization of an ancestor in the adult stage. Provided that the current opinion, that the crustaceans are derived from annelid-like forms, is correct (and there is no reason to doubt it), the body of the adult primeval crustacean was elongated, with many segments bearing extremities, and thus entirely unlike the *nauplius*.

This does not mean that the *nauplius* is entirely devoid of adult characters. It differs from the *trochophora* in two main features; firstly by the absence of cilia and secondly by the possession of three pairs of extremities. At least the second of these two features must be considered as an adult character.

These three pairs of extremities are retained in the adult, though with partly changed shape and function. They are developed very early and are ready for use when the larva hatches. Their premature development was a requisite condition for the elimination of the original organ of locomotion, the ciliation, which had been inherited from the *trochophora*. Cilia are almost completely absent in the Arthropoda, and it may be assumed that their disappearance began in the adult.

We must not of course imagine that the three extremities have suddenly appeared fully developed in the young ciliated *trochophora*-like larva, but rather that they gradually appeared at younger and younger stages. In this connection it should be noted that in the *metatrochophora* and *nectochaeta* stages of the polychaets the parapodia are more or less developed before the complete disappearance of the ciliated rings at metamorphosis. A similar process must have occurred in the evolution leading to the crustaceans. It is thus not difficult to imagine how an adult character such as the extremities might have arisen in the *nauplius* by adult pressure, and consequently how this larva was derived from the *trochophora*.

I want to stress this point because objections against the derivation of the *nauplius* from the *trochophora* have occasionally been raised. Korschelt and Heider (1936, p. 630) draw attention to the fact that in contrast to the *trochophora*, *nauplius* is segmented. This is not, however, a valid objection, since the *trochophora*, if it is not already segmented, will very soon acquire segmentation (*metatrochophora*). The early segmentation exhibited by the *nauplius* is moreover connected with the shifting forward (acceleration) of the formation of the three pairs of appendages and therefore an adult character in its own right. It should be noticed, furthermore, that the similarity with regard to segmentation and extremities is still greater between the next stages, i.e. between *metatrochophora* and *metanauplius*. It would be strange if there were no similarities between the primary larvae of annelids and crustaceans, as these groups have common ancestors.

When I maintain that the occurrence of the three extremities in *nauplius* is to be interpreted as an adult character, this is only correct in so far as the same structures are found also in the adult. Their modification with long swimming hairs, etc. which is specific to the larva has evidently only taken place after their appearance in the larva, and for this reason the modification as such is a purely larval character—an adaptation to the pelagic zone (cf. p. 223).

In fresh water ostracodes belonging to the family Cypridae another adult character is encountered, the initial development of which has been transferred to the larva. Although in other respects the larva is a typical *nauplius* the characteristic adult lateral shells are already fully developed when it hatches from the egg. This can only be explained as an effect of adult pressure. Originally the nauplii of all crustaceans must have been without shell structures of any kind (cf. the situation in the *trochophora*). In other ostracodes the change has progressed still further. In marine genera showing brooding (*Cypridina* and others) direct development has arisen, i.e. adultation has become complete.

In the Cirripedia the *metanauplius* stage has a well-differentiated shield- or dish-shaped dorsal shell which is reminiscent of the shell in the Notostraca among the phyllopods. In certain balanids this shell is particularly distinct. Since the shell is not a recent adult character, and since it can hardly be regarded as an adaptation to the pelagic mode of life, it is the logical conclusion to consider this shell as an ancient adult character.

The same interpretation can be applied to the shell in the next stage, the *cypris*-larva, where it has become differentiated into two lateral valves.

The application of the principle of adultation to the Cirripedia thus leads to the conclusion that the free-living adult ancestors of this group during some period of their evolution possessed first an undivided dorsal shell and later on two lateral valves. Since such shells and carapace structures of different kinds are still found in other groups of the crustaceans, it would be interesting to find out if and to what extent they represent homologous elements. It is also of importance to establish whether or not the shell structures of the present cirripedian adults can be traced back wholly or in part to the shells of the larva. This possibility is not excluded by the fact that these are shed, e.g. in the balanids, at metamorphosis, since regeneration may occur (cf. the tentacles of the *actinotrocha*, p. 27). These problems, however, fall outside the scope of this book, and this also applies to the problem of whether within the phylogenetical history of the present cirripedians a transition to the sedentary mode of life has taken place only once or on several occasions. The matters mentioned illustrate, however, the importance of paying attention to ancient adult characters in the larvae (see also p. 250).

A more detailed review would certainly result in the discovery of quite a number of ancient adult characters in the crustaceans. Here I shall be satisfied to give just one more example, the occurrence of an exopodite on the paraeiopods in larvae of *Decapoda*. In the adult this appendage is missing, but since it is present both in the larvae and the adults of several other groups, I find myself compelled to assume that the ancestral forms of the decapods had an exopodite in the adult stage. I imagine that the specialists on crustaceans consider this as self-evident, but it may never-theless merit pointing out, since it supplies an example of a character that has disappeared in the adult only recently.

It is out of the question to analyse here all the different larval stages which appear in the crustaceans after the *nauplius*. This would be a task in itself. I must, however, point out that in spite of the varying and in part very complicated ontogeny in the group, nothing emerges to contradict the main ideas in the present work. The difference compared with most other groups with a primary larva lies in the fact that no rapid metamor-phosis takes place. This is replaced by a sometimes considerable number of small metamorphoses, conditioned by ecdysis. These constitute the main difference between the ontogeny of the crustaceans and that of the polychaets. As is the case in many polychaets, the pelagic phase is often more or less lengthened (cf. Spionidae). It might be imagined that such a lengthening has contributed to the establishment of the holopelagic mode of life. This is, however, a complex of problems upon which I cannot enter, and which should be taken up by some specialist on crustaceans.

Even in cases where a certain degree of direct development has arisen in connection with the loss of the *nauplius* as a free larval stage and its transformation into an embryonic stage inside the egg envelope, the juvenile form which is liberated at hatching, and in many cases also the ensuing stages, is generally free-swimming even in forms which in the adult phase are pronounced bottom-dwelling animals. Such pelagic juveniles are usually still quite unlike the adult. As a rule, however, every ecdysis results in greater similarity with the adult, even if this development is complicated by a number of special larval characters (floating appendages of different kinds, etc.).

In certain cases these special characters may prevail. This is exemplified by the loricate decapods. Here, as is usual in the decapods, the *nauplius* stage is incorporated into the embryonic phase. On hatching, however, a larva is liberated which differs considerably from the adult and develops into the well-known *phyllosoma*-larva. With its flat body and long extremi-ties this larva, which with increasing age can reach an enormous size, con-stitutes a pronounced adaptation to a lengthy pelagic phase. For this reason it must be transformed at the transition to life on the bottom, undergoing a kind of metamorphosis. We therefore cannot expect the *phyllosoma*-larva to provide us with any new information about the adult of the ancestral forms. Prior to the evolution of this larva, which must be phylogenetically quite young, the post-embryonic phase of development was more direct than it is now in these decapods. The *phyllosoma*-larva must therefore be regarded almost as a secondary larva (cf. certain larvae of fishes, p. 211).

The distribution of free *nauplius* and of direct development within the crustaceans shows clearly that the transition to direct development has taken place independently in several instances during the phylogenetic history of the group. This explains why direct development exhibits so many variations, including transitional types from the development via a free *nauplius*. An example of such a transitional type is given by the cladocer *Leptodora hyalina* in which a *metanauplius* is hatched from the winter eggs, while the summer eggs develop directly.

It has already been mentioned in passing that in cases with direct development the *nauplius* is incorporated into embryogenesis and "degraded" to a stage within it. This has been established in several cases, even in forms with very yolky eggs, like *Astacus*. This feature can only be explained as being reminiscent of ancestral forms with a free *nauplius*-larva. The opposite alternative, that the crustaceans have evolved, often in parallel, towards the *nauplius*-larva with its adaptations to a pelagic existence, seems to me quite out of the question. Such evolution is not only improbable by itself, but it would also imply that among other things a rich supply of yolk in the egg and more or less complicated brooding were original features that are on their way towards elimination. Evolution would furthermore move *towards*, not *away from* total cleavage, coeloblastula, and gastrulation by invagination—all of them features that are characteristic mainly of the "lower" groups. In such a case the Malacostraca would be the most primitive group in these respects. This kind of reasoning is preposterous.

No, here as elsewhere in the metazoans evolution has certainly moved towards direct development. It is, however, very interesting to find that there still exist in the Malacostraca some cases with a free *nauplius*, for example Euphausiacea and even some members of the decapods (*Penaeus*, *Lucifer*, and others). It should be noticed that these forms differ also by the absence of brooding. In certain cases the eggs are freely discharged into the water. The genus *Lucifer* deserves particular interest because of the small amount of yolk in the eggs, coeloblastula, and gastrulation by invagination. It is remarkable that these original features (see p. 240) have been retained high up among the decapods. In their invagination *Astacus* and other genera in which the eggs are rich in yolk show how tenaciously this original mode of gastrulation can be retained in forms with otherwise strongly altered ontogeny.

Chaetognatha

This group consists of holopelagic forms apart from those which have secondarily returned to a benthic existence (*Spadella*). Direct development occurs without exception and is here so complete that no trace of the features of the once existing primary larva can be established in embryogenesis. It may be said that *total* adultation has taken place. This makes it very difficult to derive any information from the ontogeny about the systematic position of the group. Direct development is to be considered here, as in many other cases, as an effect of the holopelagic mode of life (see p. 236).

In spite of the altered life cycle the ontogeny nevertheless contains certain primitive traits: in some cases the eggs are discharged freely into the water, where they remain floating throughout development (*Sagitta*), a coeloblastula always occurs, and gastrulation is effected by invagination.

Echinodermata

The adult forms of echinoderms (with the exception of *Pelagothuria* and perhaps some ''bottom-pelagic'' holothurians) are typically benthic, and the larvae usually pelagic. According to Thorson's (1946) investigations in Öresund not less than 88–89 per cent of the species with known ontogeny have just such a pelago-benthic life cycle. These observations indicate that a life cycle of this kind is original for the echinoderms.

Other commonly-occurring primitive traits within the group are the free discharge of the sexual products into the water (which, according to Thorson, 1946, p. 366, applies to all Nordic echinoderms with pelagic larvae), the relatively small quantity of yolk in the egg, external fertilization, holoblastic cleavage of the egg, ciliated and free-swimming coeloblastula, gastrulation by invagination, and planktotrophy of the larva.

This assemblage of primitive features makes the echinoderms one of the groups which, as far as the majority of the species is concerned, in the essential aspects described have not at all progressed from the first coelomates, not even from the last common ancestor of these and the cnidarians (see p. 240). The exceptions which occur now and then with regard to the characters mentioned most of which are mutually connected, must be due partly to changes in the special lines of evolution leading to the present subgroups, partly to those within taxa of still lower rank.

The greatest concentration of divergences from the original situation is found in forms with brooding and in viviparous forms. Both kinds are found within all subgroups.

Remarkably enough the otherwise most primitive subgroup, the Crinoidea, exhibits in its development several changes which in the other subgroups are found only in a minority of species. Thus all crinoideans with known ontogeny have eggs which are rich in yolk, as well as lecithotrophic and also otherwise altered larvae. Brooding is common, and in certain cases the pelagic phase of the life cycle has been entirely eliminated, development up to the stalked stage taking place inside the mother animal.

All subgroups with the exception of the Crinoidea have a larva of the basic type, usually called *dipleurula*. It might be more correct to say that the ontogeny includes a stage of this name characterized in its external appearance by a ventral ring of cilia surrounding the depressed mouth region. During development to the final larval stage this ring is strongly extended into bights in different directions, and this is usually connected with the formation of arm processes of varying shape. This results in the development of different forms, *auricularia*, *bipinnaria*, *brachiolaria*, and *pluteus* with rather dissimilar outward appearance.*

All these different forms of the *dipleurula* type are planktotrophic and have a pelagic phase of varying length. Like other primary larvae they exhibit features which are adaptations to the pelagic zone, as, for example the various processes, and the band of cilia which has been altered in the way described.

The adaptation of the larva and the adult to their respective biotopes has resulted in considerable divergence between these two stages. This divergence doubtless occurred already in the free-living ancestors, but grew more pronounced after the adult had become sessile and consequently had acquired radial symmetry.

The increasing divergence resulted in ever more far-reaching metamorphosis. At last the transformation of the larval body was so profound

---

* In the literature the term *dipleurula* also has another implication. Bather (1900) applied it to a hypothetical ancestral form of the echinoderms which in its essential features was considered to agree with the larvae mentioned.

that in many recent forms considerable parts of the body had to be discarded. This is especially striking in certain asteroids and ophiuroids (e.g. *Luidia* and *Ophiothrix*) in which the small adult is detached from the larval body like a bud.

Another complication in the development of the echinoderms is that, in addition to the *dipleurula* type another kind of larva occurs, the so-called *doliolaria*. This is characterized mainly by circular bands of cilia, as a rule 4 or 5 in number, and lecithotrophy. It occurs in Crinoidea, where the *dipleurula* type has not been established, and in Holothuroidea. Furthermore it can be noticed that in some few cases more or less distinct *doliolaria* characters (cilia rings or indications of them) have been established in the larval development of Ophiuroidea and Echinodea (Grave, 1903; Mortensen, 1921).

Within the Holothuroidea there is considerable variation in the life cycle. In species of *Synapta* and *Holothuria* for instance a planktotrophic *auricularia* is followed by a *doliolaria* which via a so-called *pentactula*, i.e. a stage in which the five tentacles have already appeared (Fig. 49) transforms itself into the benthic adult. In many holothurians the *auricularia* stage is, however, omitted, that is to say the *doliolaria* is developed directly from a uniformly ciliated stage. This applies, for example, to *Cucumaria echinata* (Ohshima, 1921) and *Labidoplax buskii* (Nyholm, 1951). In other cases, for example, in *Cucumaria frondosa* (J. and S. Runnström, 1921) the *doliolaria* rings, too, are omitted.

In another variant a fully developed *pentactula* is hatched from the egg and immediately starts life on the bottom. There are, finally, cases of brooding and vivipary in which the liberated young resembles the adult to a still higher degree. Examples include *Chiridota rotifera* (Clark, 1910) and *Holothuria floridana* (Edwards, 1908). In the latter the young has both tentacles and feet when it hatches.

It is of particular interest that traces of the pelagic larval phase, i.e. both *auricularia* and *doliolaria* characters, can be found in such cases of a holo-benthic life cycle, and this presents a distinct parallel with the situation in, for example, holobenthic gastropods. This, in addition to the fact that in the forms without *auricularia* the eggs contain a greater quantity of yolk, clearly indicates that a life cycle of the kind found in *Synapta* (a life cycle with a planktotrophic *auricularia*) is the most primitive found in the holothurians.

Although it has the adult feature of tentacles the *pentactula* often prolongs the pelagic existence. Examples are provided by *Labidoplax* (Nyholm, 1951) and other apodous forms (own observations in tropical waters). I have found this also in two other larvae probably belonging to the genus *Cucumaria*.

In the late summer and autumn (August to October) isolated specimens of these two larvae are found quite regularly in plankton samples from the Gullmar fiord and adjoining parts of the west coast of Sweden. They are on the whole similar, but differ mainly in the colour of the anterior part of the body, one being yellowish, the other pale blue. The colour may be due to the enclosed mass of yolk.

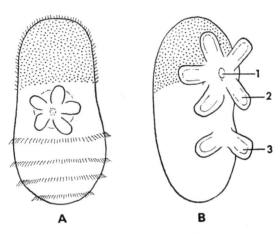

Fig. 49. Ontogenetic stages of a holothurian (probably *Cucumaria* sp.). A, transitional stage between *doliolaria* and *pentactula*; B, the first benthic stage. The dotting in the anterior end marks the distribution of a yellowish tinge. 1, mouth; 2, tentacles; 3, first pair of feet.

I want to summarize the following comments about the yellow species. Shape and general appearance of a transitional stage between *doliolaria* and *pentactula* can be seen in Fig. 49A. The vestibulum bearing the tentacles is situated approximately in the middle of the body. In the centre of the ring of tentacles a depression, the prospective mouth, can be detected. At the anterior part of the body the feebly developed ciliation is uniform. On the posterior half it is distributed on four rings (it is often difficult to distinguish more than three).

The swimming locomotion was found to be particularly interesting. The larva sometimes moves with the anterior, sometimes with the posterior end foremost, and in either case the rotation around the longi-

tudinal axis may be in either direction. Occasionally no rotation takes
place at all. If the latter case is included, we have no less than six com-
binations. Usually, however, the larva swims with the posterior end foremost, the rotation being clockwise as seen from in front. While
swimming in this manner it keeps to the bottom of the dish. Sometimes
it changes direction and moves with the anterior end foremost. In so doing
it rises in the water, and the direction of rotation is maintained. According to Nyholm (1951), the larva of *Labidoplax* can also move with either
end foremost and the same applies to the larva of Phylactolaemata (see
p. 36). The phenomenon is of particular interest in that it shows that the
orientation of the body in swimming gives no indication of which end of a
larva is anterior and which is posterior.

The phylogenetic relationship between the *dipleurula* and the *doliolaria*
type is an important question. Part of what has been said above about the
characteristics of the Holothuroidea, especially the fact that the *doliolaria*
is lecithotrophic, wherever it occurs, indicates that of the two the *di-*
pleurula type is the more primitive (cf. p. 243).

This conclusion had already been reached by some scientists by the end
of the last century, and it was adopted by the well-known Danish
specialist on echinoderms, Th. Mortensen (1921). Grave (1903), on the
other hand, maintained the opposite view, that the *doliolaria* should be
considered as more primitive. Because Grave's hypothesis has been
adopted by some later authors—e.g. Korschelt (Korschelt and Heider,
1936) is of this opinion—I am forced to scrutinize his arguments.

Grave supports his opinion by saying that the complicated metamorphosis characteristic of the planktotrophic larvae of echinoderms show
that these larvae "have been carried far out of the path of phylogeny".
By this remark Grave means to say that these larvae, the *dipleurula* type, do
not recapitulate the adult ancestral form.

The *doliolaria*-larva, however, should exhibit recapitulation. With its
successive metamorphosis it should give a truer picture of "the past
history of the race". In other words, according to Grave the original life
cycle of the echinoderms should have contained only a *doliolaria*-larva
which developed into the adult without any major changes.

As can be seen, Grave's way of reasoning is based at least in part upon
Haeckel's theory of recapitulation in its extreme form. Consequently he
considers that in the phylogeny of the present echinoderms there must
have been a stage with a pelagic adult which had rings of cilia similar to
those found in the *doliolaria*. In other words, development must have been

direct. The *dipleurula* and complicated metamorphosis would have arisen later.

As we now know quite definitely, Haeckel's theory of recapitulation is not applicable without reservations in the case of pelagic larvae (see p. 248 *et seq.*). As far as ciliation or other purely larval (pelagic) characters are concerned, neither the *dipleurula* nor the *doliolaria* can be considered as recapitulating the adult stage of any ancestral form. All the indications are (cf. above) that the long series of different ancestral forms of the echinoderms have all had a pelago-benthic life cycle.

Thus the train of thought pursued by Grave cannot lead us to an idea of the larva of the "primeval echinoderm", whether it was of *dipleurula* or of the *doliolaria* type. The fact that in most cases the *dipleurula* type undergoes far-reaching metamorphosis tells us nothing in this matter, since such a metamorphosis is a result of the divergence between larva and adult (see p. 217).

Another approach will have to be used in the attempt to arrive at a decision as to which type is the most primitive. We have already found that in cases where both types occur in one and the same life cycle the *doliolaria* always follows after the *dipleurula* and not the other way round, and that the former, wherever it occurs, is without exception lecithotrophic. It must be added that another group among the deuterostomians, the Enteropneusta, has a larva, the *tornaria*, which agrees closely with the *dipleurula* in the very special configuration of the ciliation. All three facts indicate that the *dipleurula*-larva is the most original.

The *doliolaria* is a lecithotrophic second stage in ontogeny, a "pupal stage", as is the case with the *cypris*-larva of the cirripedians. In many of the life cycles of the echinoderms the first stage, the *dipleurula*, however, has been lost in connection with an increased amount of yolk in the egg and brooding of different types.

The question now is where in phylogeny this second stage had its origin. Because lecithotrophic and otherwise secondarily altered larvae of enteropneusts (see p. 198 *et seq.*) do not exhibit any *doliolaria* features, it seems most likely that these features did not arise until after the echinoderms had split off as a branch of the genealogical tree (see Fig. 57).

It is difficult to say whether the life cycle of the "primeval echinoderm" already contained a *doliolaria* stage which was later eliminated in some of the evolutionary lines leading to the present forms, or whether the group has a general tendency to produce *doliolaria* features in connection with the change to lecithotrophy. The fact that in certain other groups one and the same change in the original larva has taken place independently in connection with the transition to lecithotrophy (see the bryozoans, p. 34) might speak in favour of the second alternative.

It is, however, worthy of notice that no *doliolaria* characters have been established within the Asteroidea, where lecithotrophy is likewise found in a number of cases. In the other subgroups, too, here and there strongly altered, lecithotrophic larvae without *doliolaria* characters are encountered.

The lack of uniformity in the number of circular ciliated bands might support the alternative that they have appeared in the echinoderms on different occasions. The usual number is four or five of which three are situated on the part behind the vestibulum (e.g. *Antedon*). In several holothurians, however, all four or five rings have been found in this position (Fig. 49).

In this connection it is of interest to note that circular ciliated bands of a secondary nature have evolved in older larvae elsewhere in the animal kingdom. Examples are *Chaetopterus* in the Polychaeta and gymnosomatous pteropods in the Gastropoda.

It thus seems reasonable to assume that the *doliolaria*-larva has arisen independently several times within the echinoderms. Fell (1948) has also arrived at the same conclusion, although for other reasons.

I am, however, unable to share Fell's opinion that the larval forms of the echinoderms, being "specialized stages", are without any great phylogenetic significance. Fell writes (p. 104): "Consequently, since we cannot interpret resemblance between the auricularia and the bipinnaria as indicating any close relationship between the two classes which possess these larvae, neither can we attach any greater importance to the resemblance between the same auricularia and the tornaria of hemichordates." From this Fell draws the conclusion "that hemichordates do not exhibit any significant relationship with echinoderms". I quite agree that Holothuroidea and Asteroidea need not necessarily be considered as more closely related to each other than to the other groups of the echinoderms for the simple reason that they have similar larvae. I have already arrived at the conclusion that the *auricularia* and the *bipinnaria* (this applies of course to the simpler forms) are closest to the original larva of the echinoderms, but because primitive characters are of no importance in assessing the relationships within a group, the larvae in this case tell no more than what is already known from other evidence—that both holothurians and starfishes are groups of echinoderms. The similarity between the more primitive larvae of the echinoderms and the *tornaria* of the enteropneusts is on the other hand of particular phylogenetic importance because it is the only really obvious similarity between these otherwise unlike phyla.

Neither in this case do the larvae indicate *how close* the relationship is, but only that there *is* a relationship, and this is valuable enough. From this similarity between the larvae (in conjunction with other facts) we have already been led to another important conclusion in that *dipleurula* and *tornaria* are the most primitive larval types in the respective phyla. There is nothing to indicate that the similarity between these types is due to convergent evolution.

I should not omit to point out that Fell considers the similarity between the *echinopluteus* of the sea-urchins and the *ophiopluteus* of the ophiuroids to be the result of convergent evolution. By way of justification he states that for adult-morphological and palaeontological reasons these two groups cannot be more closely related. According to Fell's phylogenetic diagram the ophiuroids are derived from the same branch as the asteroids, while the echinoids originate from the pelmatozoans in a different way. If this could be demonstrated, then one could not object to Fell's hypothesis. But—the connection between the groups has by no means been elucidated. Fell says himself that his genealogical scheme is "in no way intended to represent established fact". In view of the present incomplete state of our knowledge I should like to say that it is just this similarity between their larvae which constitutes perhaps the most important argument in favour of a common origin for ophiuroids and sea-urchins. But it is obvious that this argument will have to give way if more convincing contrary ones can be put forward.

Because of the great changes which the adult of the echinoderms has undergone (in consequence of its sedentary mode of life) it would be of great value in continued attempts at unravelling the phylogeny of the group, if the retention of ancient adult characters dating from the very early evolutionary period when the adult was still free-living could be established in the larvae. Apart from the bilateral symmetry and the organization of the coelomic system (especially its division into three sections) no such characters seem to be retained.

In the *brachiolaria* of the asteroids on the other hand, there is a structure which must be interpreted as an ancient adult character of later date— the adhesive pit. This structure, which is situated ventrally at the anterior end, is of some interest in considerations of phylogeny.

The adhesive pit corresponds directly in situation, shape, and function with the adhesive organ in the *doliolaria* of the crinoids, and it is tempting to follow some earlier authors and think of a homology. In function there is also agreement with the adhesive organ in, for example, the larvae of the tentaculates, especially the bryozoans (see p. 34), but in this case it is evidently only a pure analogy (note the postoral position of the organ in the tentaculate larvae).

When designating the adhesive pit as an adult character I am expressing myself in a way which is in some measure inappropriate. The pit is of course only the basal portion of the stalk (i.e. the altered anterior end of the body) by which the adult was, or still is (stalked crinoids), attached. As is known a stalk is also formed in the *brachiolaria*-larva which is sedentary during metamorphosis. Thus it is more properly the whole stalk structure which is the ancient adult character.

The literature contains a good deal of discussion about what is original and what is altered in the case of the larvae of echinoderms. With what has just been said and on the basis of the fundamental theses laid down in this book I agree unreservedly with the opinion that with regard to the possession of the adhesive pit and the stalk the *brachiolaria* is more primitive than are larvae (e.g. the *bipinnaria*) which lack these structures and metamorphose in the pelagic zone. At one time all echinoderms were sedentary, and at metamorphosis the preoral part of the larva gave rise to a more or less stalk-like structure. Considering the usual effect of adult pressure it appears to me most unlikely that the preoral part of the ancient pelagic larvae should have undergone no modification at all. At least some adhesive gland cells should have been found, but it appears more probable that at one time all echinoderm larvae possessed an adhesive pit of a type resembling that in the *brachiolaria* and the *doliolaria* of the crinoids. When the adults of all those echinoderms which are no longer sessile in any phase of their life cycle had again become free-living, their larvae were no longer in need of an adhesive organ, and for this reason it has been eliminated in most of them. Its retention in the *brachiolaria* is a phenomenon of delay. As far as ancient adult characters in the larvae are concerned, they are only remnants. Such characters are always on their way towards elimination, even if the process is slow.

I can thus not share the opinion of Mortensen and certain other authors that the adhesive pit of the *brachiolaria* is a new formation which has nothing to do with the corresponding organ in the larvae of the crinoids. With regard to the possession of the adhesive pit the *brachiolaria* is more primitive than the *bipinnaria* and other larval types in which this structure is missing.

In the above discussion about the adhesive pit of the *brachiolaria* I have intentionally disregarded the three special arms situated around the pit which also serve for attachment. These "*brachiolaria*-arms" are without any counterpart in other echinoderm larvae. For this reason they give the impression of being entirely new formations (accessory adhesive organs) within the Asteroidea.

It appears, however, that the adhesive arms of the *brachiolaria* should be interpreted somewhat differently. These arms differ from the other arms of the larva not only by the fact that they have adhesive cells at their tips, but also by containing a diverticulum of the coelom from the axohydro-coel. Thus they agree both in structure and function remarkably well with the ambulacral feet of the adult. (The similarity in function refers not only to attachment but also, at least in certain forms, to locomotion.)

Is this similarity purely accidental? I do not believe so. In view of the principle of adultation we have to reckon with the possibility that the adhesive arms are an adult character that has been transferred to the larva. These arms might be described as ambulacral feet, even if they have been modified.

I will readily admit that this attempt at interpretation may appear bold and provocative. But a transfer of this kind is not entirely unique within the phylum. At least two other polymerous structures which are charac-teristic of the adult may appear in the pelagic larva; these are pedicellaria in *echinopluteus* and "calcareous wheels" in the *auricularia* of certain holothurians. There is really no difference other than that in the *brachio-laria* the adhesive arms have a function in metamorphosis; this is not the case with the other structures mentioned.

It would naturally be interesting if it could be convincingly determined where in the evolutionary tree of the echinoderms the adhesive arms have made their appearance, i.e. if some opinion could be formed about the phylogeny of the *brachiolaria*. Unfortunately not much can be said with certainty on this point. I have already arrived at the conclusion that at metamorphosis the larva of the primeval echinoderm attached itself with the anterior end by means of the counterpart of the present adhesive pit, and that as ontogeny continued this end was lengthened into a stalk. There is nothing to indicate, however, that the adhesive arms were al-ready present at this evolutionary stage. If they are adult characters, as discussed above, their occurrence in the larva of the primeval echinoderm is not feasible for the simple reason that its (sessile) adult did not possess any papillae designed for locomotion (ambulacral feet). A locomotory function of the papillae evidently arose only after the branching-off of the evolutionary line leading to the recent crinoids. In this group the papillae have never functioned as organs of attachment or locomotion, and for this reason it is not to be expected that the larvae of the crinoids and their ancestors possessed any adhesive arms.

Concerning the other recent echinoderm groups it can only be said that the emergence of the adhesive arms in the larvae may have taken place anywhere in this part of the phylogenetic tree after the adults had acquired ambulacral feet. (In those groups which now are without adhesive arms, these may have disappeared.) It seems, however, most likely that the emergence of the arms took place only in the special branch leading to the Asteroidea. In this group they are found, as far as is known, in all forms except *Astropecten* and *Luidia*. It is difficult to say whether or not their absence in these two genera is a primary feature, but because their *bipinnaria* had certainly lost the adhesive pit, it might seem that the lack of adhesive arms is likewise a secondary feature. At all events it is possible to imagine that the loss of the two components of the adhesive apparatus took place more or less at the same time, "shortly" after the larvae had ceased to attach themselves. But I want to stress once more that there are no definite indications when the emergence and disappearance of the adhesive arms took place in the phylogeny of the echinoderms. (This uncertainty occurs irrespective of whether or not they are to be considered as an adult character.)

Some further remarks about changes in the life cycle of the Echinodermata might be added. In connection with a large quantity of yolk in the egg very marked transformations of the larva have occurred at different instances within the phylum. In many cases the larva has become lecithotrophic and considerably simplified in its external appearance. The body is frequently without such differentiation as arm processes or bands of cilia which characterize the original (planktotrophic) larva. Very often, and this applies mainly to holothurians and starfishes, the entire surface of the body bears a uniform ciliation which is evidently an embryonic relic. (N.B. the blastula and gastrula stages of forms with planktotrophic *dipleurula*-larvae usually have such ciliation!)

As indicated by the ciliation these larvae are often still pelagic. The

Fig. 50. Embryo of *Asterina exigua*, in different views. In this species the life cycle has become holobenthic. Of the external characters of a typical *brachiolaria*-larva only the adhesive arms and the adhesive pit are still retained. In other star-fishes larvae of this type (though usually with smaller arms) may be still pelagic. (After Mortensen, 1921.)

pelagic phase may, however, be entirely eliminated (Fig. 50). This is especially common where brooding on or within the mother animal has arisen. Development is then more or less direct. For a still unnamed, oviparous ophiuroid, Fell (1941) has described a fully perfected direct development. Here the newly hatched young is a pentameric creature which even during embryogenesis does not exhibit any trace of pelagic larval characters (Fig. 51). In this form recent adult pressure—adult

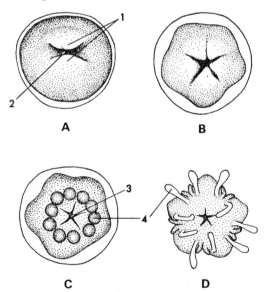

Fig. 51. Ontogenetic stages of an egg-laying ophiuroid with direct development (name of species not given). A–C, embryonic stages within the egg membrane; D, newly hatched young. 1, "epibolic crests"; 2, blastopore; 3, mouth; 4, ambulacral feet. (After Fell, 1941.)

pressure that has arisen after the echinoderms became radially symmetrical in the adult stage—has taken complete effect. (A last trace of the original bilateral symmetry might perhaps be interpreted in the fact that the gastrula is not perfectly radial, but has two "epibolic crests", Fig. 51A.) Another interesting feature of this form is that the mouth is formed, if not directly from the blastopore, at least in its place, constituting part of the considerable alteration in ontogeny which this species has undergone.

Pogonophora

In this peculiar group of animals, which are without an alimentary tract in the adult, much remains to be elucidated about the larva and its development. It is known, however, that the pelago-benthic life cycle is

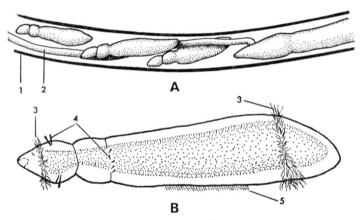

FIG. 52. *Siboglinum* sp. A, three larvae ready to leave tube of mother animal. B, such a larva under stronger magnification (*ca* × 100), entoderm indicated by dotting. 1, wall of tube; 2, tentacle of mother animal; 3, ring of cilia; 4, chaetae; 5, ventral band of cilia. (After Jägersten, 1957.)

represented, even if probably the larva swims about only for a very short time. Investigations by Jägersten (1957), Ivanov (1963, and earlier publications), and Webb (1964) have shown that the larva is lecithotrophic. Its appearance is represented in Fig. 52B. The most obvious external characters are two strong rings of cilia, a longitudinal, and probably (see below) ventral band of cilia, and chaetae of two different kinds. No trace

of mouth or anus can be detected, but in contrast to the adult the ento-
dermal part of the alimentary tract is retained. This mesenteron is, however, without a lumen and it functions only for the storage of the major
part of the mass of yolk derived from the egg.

The above proves that this larva is considerably altered. The absence of
mouth and anus might be interpreted as a result of "adult pressure",
since they are also absent in the adult. It must, however, be kept in mind
that the openings may be reduced or even entirely eliminated in lecithotrophic larvae of groups which have a complete alimentary tract.

Obviously only the two rings of cilia are purely larval characters—
adaptations to the pelagic life—which are lost during the transformation
into the adult. Direct observations prove that the rings of cilia really are
organs of locomotion (Webb, 1964; see also Ivanov, 1963, p. 117).
Ivanov assumes that the posterior ring is homologous with the circular
band of cilia in the tornaria-larva. This might appear probable, but since
it has been established that the ciliation of lecithotrophic larvae may have
been transformed into rings this interpretation is somewhat uncertain
(note the doliolaria of the echinoderms). The appearance of the once
planktotrophic larva of the ancestral pogonophores is also uncertain, but
for the reasons which apply also to the chordates (see p. 201) it may be
assumed that it was of the dipleurula type. I thus find it not impossible that
the larva of the pogonophores might have been altered relatively "recently" in certain parts of its ciliation. But I do not believe that this
applies to the band of cilia situated between the two rings.

This longitudinal band seems to be identical with that in the adult
situated ventrally on the anterior part of the metasome (for the division
into regions, see below). For this reason it should be interpreted as a
recent adult character. My examinations of Siboglinum (Jägersten, 1956,
1957) lead me to think that the band in question is a relic of the ventral
ciliation used by the free-living ancestral forms of the pogonophora for
gliding locomotion on the substratum.

My comparative study of the larvae of the metazoans has led me to a
deepening conviction about the correctness of this concept. The band is
obviously a counterpart of the ventrally situated band of cilia which is
found not only in the relatively closely related tornaria-larva (see p. 199),
but also in the larvae of brachiopods, annelids, molluscs, entoprocts, and
sipunculids (see these groups: see also p. 244). Apart from certain exceptions (e.g. Protodrilus and other archiannelids) this ventral band of cilia
is no longer found in the adult, but its occurrence in the larva shows that

it once existed in the ancestral forms and functioned as an organ of loco-motion. It is of special interest that this must apply also to the now sedentary groups Brachiopoda, Entoprocta, Sipunculida, and Pogonophora.

It appears from this discussion, that I adhere to my previous opinion about what is ventral and what is dorsal in the pogonophores. A more de-tailed discussion of this controversial question falls outside the scope of this book. I must, however, point out that my opinion is also supported by other facts. Thus attention should be paid to the fact that the great longitudinal nerve trunks are situated on the same side as the band of cilia, a feature found in all groups where both structures occur, *and that in all of them this side is ventral*. In other words, the side with the longitudinal band of cilia in the pogonophores corresponds to the ventral side of the other groups of invertebrates.

The reasoning causing Ivanov, Webb, and others to consider the side with the band of cilia as the dorsal side is obviously a one-sided com-parison with the chordates. This leads to the erroneous conclusion that the body of the pogonophores is, so to speak, on its back, and in this way all connection with the other invertebrates is lost.

A comparison between the pogonophores and the chordates, however, is still of interest, but *then the body of the chordate must be placed on its back instead*! In this case the chordates seem to be the altered group. The orientation of the chordate body in this way brings to light a remarkable degree of agreement, *inter alia* in the location of the great nerve trunk, the approximate position of the central pumping organ for the circulatory system (''the heart'') and also the direction of the blood-flow. Here I am touching on an extremely important question concerning the phylogeny of the chordates, which I hope to treat in greater detail in a future paper.

Something further must still be said about the ventral band of cilia in the pogonophores. Earlier investigations (1957) have led me to the con-clusion that it is the same structure both in the larva and in the adult. Webb (1964) is of the same opinion. Ivanov (1963 and earlier papers) who believes that the band of the larva is situated on the mesosome is, on the other hand, prevented for this reason from accepting the view of Webb and myself. According to Ivanov the band of the larva should be lost and a new one formed later on the metasome of the adult. Although at the present stage of our knowledge this appears hardly probable, the essential point would hardly be altered, even if Ivanov contrary to all likelihood is right, the fundamental principle being that we are concerned with a remnant of the ventral organ of locomotion in the non-sedentary ancestral forms, although different parts of it in this case.*

The chaetae are structures which, like the ventral band of cilia, should probably be considered as an adult feature. Here no more than in the annelids can it originally have been a question of an adaptation to the pelagic mode of life. For this reason the occurrence of chaetae in the larva should in all probability be considered as a result of "adult pressure".

We have already seen indications that the pogonophores are a very strongly altered group. This applies to the gametes as well as to all other phases of development: egg cells rich in yolk, sperm which differ from the primeval type, occurrence of spermatophores (and in connection with this probably internal fertilization), narrowed blastcoel, divergent gastrulation (i.e. no invagination), brooding, and lecithotrophic and also otherwise altered larvae. It would hardly be inaccurate to say that development is on the verge of direct development. For this reason it is all the more interesting that there are still purely larval (pelagic) characters (the rings of cilia), and that the larva still swims about freely, albeit for a very short time and near the bottom. The retention of these remnants of more primitive features is probably necessary for the transition from the tube of the mother animal to a place suitable for continued development.

Enteropneusta

Like the majority of the relatively closely-related echinoderms the enteropneusts generally have a pelago-benthic life cycle and a primary larva of *dipleurula* type—the *tornaria*.

The enteropneusts are almost less primitive than the echinoderms in the earlier phases of development. The eggs are frequently not discharged freely into the water, but are held together by a slime (which seems, however, to be quickly dissolved by the water). Furthermore, in many

* If here and there I make use of the terms "mesosome" and "metasome", these terms are applied in accordance with the original concept of the division of the adult into regions. Webb (1964) has, however, cast new light on this question and has perhaps arrived at the correct solution. Further investigations, however, are required. Webb (1965) has furthermore arrived at the conclusion that at the transformation into the adult a minor part of the ciliated band of the larva is lost. This of course does not prevent the view maintained by me here.

cases hatching takes place only after gastrulation. Otherwise the usual primitive conditions hold: external fertilization, holoblastic cleavage, coeloblastula, and gastrulation by invagination.

The *tornaria*-larva has a long pelagic life during which it frequently reaches a remarkable size, specimens with a length of considerably more than 1 cm having been encountered. The larva is of course planktotrophic, its food consisting of small particles which by a complicated system of ciliated grooves are carried to the mouth (see Garstang, 1939).

First of all, because of its close morphological agreement with the *dipleurula*-larvae of the echinoderms the *tornaria* must be considered as the most primitive larval form within the Enteropneusta. This implies that the so far little-examined larvae in the subgroup Pterobranchia are altered forms which have evolved from the *dipleurula* type. This leads also to the conclusion that the alleged similarity with (likewise altered) larvae of bryozoans is simply a phenomenon of convergence. It is important to note that the known larvae of Pterobranchia (they all belong to the family Cephalodiscidae) are lecithotrophic. Here as in all metazoans, with the probable exception of the Spongiaria, lecithotrophy is a secondary phenomenon (see p. 224 *et seq.*).

Since these larvae are often compared with those of bryozoans it may be important to examine the alleged similarity more closely. Fig. 53A shows that it is mainly the occurrence and location of two organs—a pre-oral glandular region and a postoral adhesive organ, which are responsible for a certain degree of agreement with a lecithotrophic bryozoan larva (cf. Fig. 4B). The glandular region should correspond with the pyriform organ, and the adhesive organ with the organ of attachment in the larvae of the bryozoans.

The external similarity between the larva of *Cephalodiscus indicus* and the lecithotrophic bryozoan larva is so great that Hyman (1959) suspects that Schepotieff (1909) had mistaken one for the other. According to the information about the development of the coelom in the former no such mistake, however, can have been committed. The development of the coelom in Schepotieff's larva shows distinctly that the larva is of an altered *dipleurula* type.

As far as can be judged from the available investigations the glandular region of the larva of *Cephalodiscus* must be transformed in the adult into the ventral surface of the cephalic shield, while the adhesive organ is transformed into the sucking disc at the end of the stalk. In the larva the two organs in question thus are evidently rudiments (anlagen) of recent adult characters. I have arrived earlier (p. 44) at the conclusion that in the bryozoan larvae also the adhesive organ (internal sac) is an adult character. The evidence thus indicates that it is not the larvae of these groups of animals that should be compared, but rather their adults. Moreover, it is very easy to arrive at erroneous conclusions if only one subgroup (Bryozoa) within a major systematic complex (Tentaculata) is compared with one subgroup (Pterobranchia) within another complex

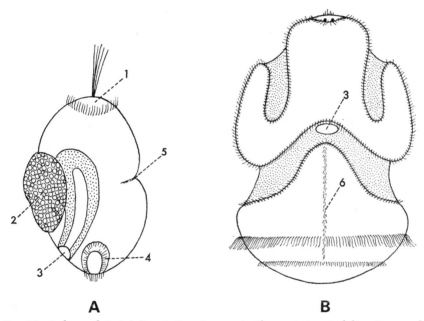

A   **B**

FIG. 53. A, larva of *Cephalodiscus indicus*. B, *tornaria* (diagram). 1, apical disc; 2, preoral glangular region; 3, mouth; 4, sucking cup; 5, position of the future anus; 6, ventral band of cilia (not developed in all forms). (A, after Schepotieff, 1909.)

(Enteropneusta). All features within the two complexes have to be taken into consideration. A task of this kind is, however, outside the scope of this book. As far as the larvae are concerned everything speaks, however, in favour of the concept that in the Bryozoa the *cyphonautes* is the most primitive larval form, while the *tornaria* is most primitive in the Enteropneusta. Thus these are the two larvae to be compared, and they are not particularly alike. It would not be justified to refer to possible similarities between altered lecithotrophic larvae within either group as evidence for a relationship between bryozoans and pterobranchians.

However, I have arrived earlier (p. 43) at the conclusion that the pyriform organ of *cyphonautes* points to the existence of a preoral creeping organ in the free-living ancestral forms of the tentaculates. There is no doubt that in the "acorn" and the cephalic shield the enteropneusts also possess an organ of locomotion in this location. We have here a similarity which might perhaps be adduced by the advocates of the view that there is a relatively close relationship between the bryozoans and the enteropneusts. Here it must, however, be pointed out that it is not in the first place the bryozoans that are concerned here, but the free-living ancestral forms of the tentaculates. It is likewise of importance that the creeping organ of these forms must have been only a part of the protosome, while in the enteropneusts the organs mentioned are identical with the whole protosome. For this reason the similarity is probably only due to convergent evolution.

Lecithotrophic and morphologically altered larvae occur not only in the pterobranchians but also in a number of helminthomorphic enteropneusts. Such a larva is known, for example, in *Saccoglossus horsti*. It is ciliated all over the body, but agrees with a typical *tornaria* only by a strong ring of cilia at the posterior end (Fig. 54A). The pelagic phase is short and lasts only one or a few days (Burdon-Jones, 1952). After the transition to the benthic mode of life a gradual change into the adult takes place (Fig. 54B and C).

In e.g. *Saccoglossus kowalevskyi* a further step has been taken towards direct development in so far as hatching takes place only after the differentiation of the three main regions of the adult body. This is in agreement with the fact that the larvae can swim only for short distances along the bottom (Colvin and Colvin, 1950). In this case the life cycle has thus become almost holobenthic and direct. The ring of cilia at the posterior end is, however, still retained as a reminiscence of an earlier evolutionary period with a *tornaria*. In other species, for example *Saccoglossus otagoensis*, even this ring has disappeared, and the pelagic phase of the life cycle is completely eliminated (Kirk, 1938).

Among the external characters of the *tornaria* the bands of cilia are very prominent. This applies especially to the often extremely complicated, labyrinth-like configuration of the two anterior bands. These bands of cilia and the tentacle-like processes which are found along them in many forms are purely larval structures that are lost at metamorphosis.

In the *tornaria* no recent external adult characters of a special kind are developed. Strangely enough there appears on the other hand in certain forms a structure which must be interpreted as an ancient adult character. This is a ventrally situated, median, longitudinal band of cilia which, at least in certain stages, extends from the mouth in a posterior direction. It passes through a small discontinuity in the anterior circular ring of cilia and reaches as far as the so-called secondary ring (Fig. 53B). This band which was observed already by Spengel (1893) and which I myself have

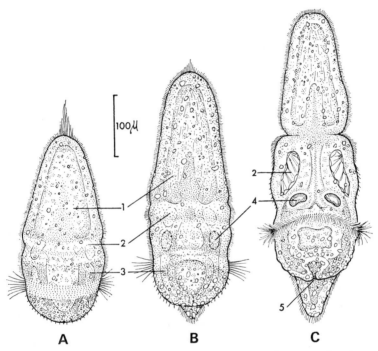

FIG. 54. *Saccoglossus horsti.* Three ontogentic stages in optical section, dorsal view. A, pelagic larva; B and C, stages shortly after the transition to life on the bottom. 1, coelom of proboscis; 2, coelom of collar; 3, coelom of trunk; 4, first gill pore; 5, anus. (After Burdon-Jones, 1952.)

found in a couple of different larvae from the Florida current agrees in position and nature very closely with the so-called neurotroch which often occurs in the larvae of annelids and certain other groups (see p. 171; see also p. 244). I must admit that I have had no opportunity to see whether the *tornaria* in accordance with other larvae creeps on the substratum by means of this band, but I have been able to observe that small particles which come into contact with it are transported backwards by its cilia. I consider it likely that here as in the annelids, etc. the band indicates that in an earlier phase of evolution the adult moved over a solid substratum by means of a ventral band of cilia rather than burrowing in the mud as it does now.

In spite of the adaptations of the *tornaria* to the pelagic zone (often implying a pelagic life of long duration and a correspondingly large size of the body) metamorphosis is still gradual. Thus nothing is known about a shedding of larval parts in the way shown by many larvae of the echinoderms. In addition to the loss of the bands of cilia and possibly existing tentacles, metamorphosis consists mainly of a shrinking of the body and the development of the two circular constrictions which in the adult separate the three regions of the body.

Finally we may recall the high degree of agreement which exists between the *tornaria* of the enteropneusts and the original *dipleurula*-larva of the echinoderms, and which is currently and in my opinion correctly considered as indicating the common origin of the two groups. (Concerning Fell's sceptical attitude in this respect, see p. 186 *et seq.*)

Chordata

In the chordates the life cycle is considerably altered. In the Asciidiacea there is admittedly a pronounced pelago-benthic life cycle, but this is obviously no longer the original one, the benthic occurrence of the adult ascidians being in all probability secondary (see below). It must, furthermore, be pointed out that with one single exception every trace of the once existing primary larva has disappeared. This exception refers to *Amphioxus* (see also p. 210). Here reminiscences of special features of the primary larva might be interpreted from the fact that the young larva is ciliated and like the majority of primary larvae swims about by means of its ciliation with simultaneous rotation around its longitudinal axis. There is thus no doubt that the ancestors of *Amphioxus* possessed a primary larva (see also below).

Upon the basis of this evidence, but also considering the relationships of the group, there can be no doubt that at one time the special line of evolution leading to the chordates included a pelago-benthic life cycle of the same kind as the one still occurring in the vast majority of other phyla. Among these are Enteropneusta, Pogonophora, and Echinodermata which

are currently supposed to be relatively close to the chordates, and which
are together with them (and usually the Chaetognatha) united as Deuterostomia.

In the discussions of the phylogeny of the chordates the tailed larva of
the ascidians has occupied a central position (Fig. 55). In later years a
number of scientists including Garstang (1928b), Berrill (1955), Whitear
(1957), and Bone (1960) have expressed the opinion that a larva of this
type gave rise to the chordates by neoteny. These authors, however, do

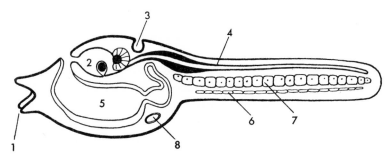

FIG. 55. Diagram of an ascidian larva. 1, adhesive papillae; 2, brain vesicle with static
organ and eye; 3, cloaca; 4, neural tube; 5, pharynx; 6, rest of caudal intestine; 7,
notochord; 8, heart.

not agree with each other on the matter of the more detailed organization
and transformation of the ancestral larva, and, primarily about the
evolutionary stage at which neoteny should have taken place. I consider it
superfluous to enter into a more detailed discussion of the different
opinions, the more so as I find myself compelled to reject summarily the
hypothesis of neoteny as far as the origin of the chordates is concerned
(see below).

Not much can be said about the features of the primary larva in the
special evolutionary line which led to the chordates. The only certain
point is that it moved by means of cilia. Since planktotrophic larvae of the
dipleurula type are found in the two most closely related groups, Enteropneusta and Echinodermata, it might, however, be justifiable to assume
that it was a larva of this kind. The problem is concerned with the way in
which the phylogeny of the deuterostomians is to be imagined. If it
followed the course which I have represented in Fig. 57 (p. 214) and
which is that currently adopted, it becomes almost inevitable to assume

that the original larva of the chordates was a *dipleurula*. (For the implication of this term, see p. 181.)

Thus I arrive in another way at the same concept as Garstang about the
original larva of the chordates. (I might point out that elsewhere I have
rejected an opinion that the *dipleurula* is without taxonomic–phylogenetic
significance; see p. 186.)

With regard to the adult in the evolutionary line leading to the primeval
chordate my concept is, however, entirely unlike Garstang's. According
to my fundamental idea about the evolution of the life cycles (see the
Introduction) a non-sedentary benthic mode of life in the adult is original
for all groups of metazoans. This mode of life had already arisen in the
Blastaea-like ancestral forms (Jägersten, 1955), and there is no reason to
assume that in this respect any change has taken place in the line leading
to the last common ancestral form of the three groups of chordates (see
phylogenetic diagram, p. 214). In the groups where the sedentary mode
of life occurs this has in all probability arisen independently in several
instances, i.e. in the special lines leading to these groups. Remane (1963)
has pointed out that the sedentary groups which Garstang and others
consider as ancestral (Pterobranchia and Tunicata) are attached by quite
different parts of the body (the posterior region of the body and the
anterior end, respectively). A course of evolution such as:

Pterobranchia → primitive forms of tunicates → other chordates

is consequently improbable for this reason. If, as suggested by certain
features, the dorsal side of the tunicates and the other chordates corresponds with the ventral side of the enteropneusts and the other
invertebrates, Garstang's interpretation becomes quite impossible.
(Unfortunately I cannot deal in this book with the possible dorso-ventral
inversion of the chordates; see, however, p. 194.)

In refuting the idea about sedentary ancestral forms the most important
argument for the hypothesis of neoteny is invalidated.

It can be assumed that the benthic mode of life in the adult was retained in the evolutionary line leading to the Chordata for some time after
the splitting-off of the line to the Enteropneusta. "Soon" afterwards a
very essential change, the transition to the holopelagic life cycle, must
have been initiated.

In my opinion this transition did not, however, take place by neoteny,
as I have pointed out above. General experience about changes in the life
cycles, and the agreement between helminthomorphic enteropneusts and
chordates over the organization of the *adult*, both argue against the

hypothesis of neoteny. I shall, however, not enter more closely into a morphological comparison between enteropneusts and chordates. This must be done in connection with the task of elucidating the evolutionary tree of the deuterostomians (I hope to return to this on a future occasion). I shall only say here that in my opinion the holopelagic mode of life has probably arisen by gradual changes and ascension of the adult into the pelagic zone in some early series of chordate ancestors some time after the evolutionary lines of enteropneusts and chordates had separated (Fig. 57). Thus in the chordates the holopelagic life cycle has probably arisen in the same way as in certain gastropods (see p. 130).

In whatever way the holopelagic mode of life arose, the primary larva, which probably was of the *dipleurula* type as suggested above, must have persisted in a more or less unaltered condition during some period of evolution. However, as, for instance in Ctenophora and Chaetognatha and in accordance with the general rule in holopelagic forms (see p. 236), "adult pressure" must have initiated the ontogenetic emergence of adult characters in the larva. These changes together with the disappearance of the special larval characters resulted finally in direct development.

The adult characters which may first be mentioned are: gill slits, neural tube, notochord, lateral musculature, and the shape of the posterior part of the body as a tail for swimming. Because the musculature took over locomotion the bands of cilia characteristic of the *dipleurula*-larva became superfluous and were eliminated.

What I have proposed here agrees with the idea of Garstang (1928b) and his followers in so far as I consider it probable that it was a larva of the *dipleurula* type which became altered. But this must not overshadow the fundamental difference in our concepts. According to Garstang the emergence in the pelagic *auricularia*-larva of certain chordate features such as neural tube, lateral musculature, shaping of the posterior part of the body as a tail for swimming, and probably also the notochord, was an *entirely new* acquisition of features to the life cycle. In my opinion it was on the contrary a matter of a transfer to the larva of the development of *features which already previously existed in the adult phase*, and the transfer was a consequence of the phenomenon which is general in the metazoans —"adult pressure".

Another important element of my concept is that the chordates are not derived from the echinoderms, either directly or indirectly. The echinoderms form a side branch of the evolutionary line which led to the chordates, among others (Fig. 57, p. 214). The *dipleurula*-larva is not restricted to this echinoderm branch.

By my discussion Garstang's derivation of the neural tube from the bands of cilia of the *dipleurula*-larva is so to speak automatically eliminated. This derivation of the adult central nervous system from a *larval* feature which is an adaptation to pelagic life (pelagic organ of locomotion) is further in itself extremely unlikely. (It would make very little difference to these considerations if there were a strand of nervous tissue underneath the band of cilia.) The central nervous system of the chordates must be derived from the central system (or part of it) of some ancestral form in its adult stage. (In a future paper I hope to return to this problem.)

The direct development resulting from the transfer of the initial development of the above-mentioned adult characters into the larval stage and the loss of the original larval pelagic features, may have become almost complete before the separation of the evolutionary lines leading to the three chordate groups. In other words, direct development might have been almost perfected in their last common ancestor, the "primeval chordate".

The situation in *Amphioxus* indicates, however, that general ciliation was found also in the newly hatched young of the primeval chordate. In other words, in the primeval chordate direct development cannot have been more accomplished than it now is in *Amphioxus*, but rather less so. Direct development without any trace of the ciliary features in the primary larva arose first in the special evolutionary lines leading to Tunicata and Vertebrata.* It can by no means be excluded, although there is no direct evidence in favour of it, that the *dipleurula*-larva might have been retained for some time in the special evolutionary lines leading to the three chordate groups, and that direct development has subsequently arisen independently in these different instances. In view of what we know about the constant tendency of the life cycle to change in this direction I consider this quite possible.

The gill slits, neural tube, notochord, and tail for swimming, which for a considerable time (as long as the primary larva remained unaltered) were restricted to the adult phase of the life cycle, are the main characters of the chordates, even if the gill slits at least, being found in the Enteropneusta, are older than the chordate type.

---

* However, in certain vertebrates cilia are still found on the epidermis of the embryo (see footnote on p. 210).

Attention must be paid to the fact that I have excluded the extremely important feature represented by myomery which characterizes both Acrania and Vertebrata to such a high degree. With the exception of these two chordate groups all deuterostomians lack myomery, and for this reason it is probable that this feature evolved only after the differentiation of the line leading to the chordates. Myomery is, however, also absent in the tunicates and their tailed larva, and this might give the impression that it arose after the branching-off of the evolutionary line leading to the Tunicata, in the line which is common to Acrania and Vertebrata.

The question is complicated by the fact that in the appendicularians the neural tube contains a series of ganglia, and that each of these sends out a dorsal and a ventral nerve which innervates the lateral muscle bands. Certain authors have interpreted these conditions as pointing to a meta-mery which was better accomplished in the ancestral forms. This concept may be correct. On the basis of our present knowledge I can express no definite opinion about the emergence of myomery and neuromery.

It can, however, be assumed with greater certainty that the primeval chordate (at least its adult) showed repetition of some other organ systems. Some degree of branchiomery certainly existed, and since the helminthomorphic enteropneusts exhibit a repetition of other organ systems in the trunk region (gonads, liver diverticula) as well as an annulation of the body wall, it seems likely that the adult of the primeval chordate, even if it was without myomery, had a more advanced segmentation than the tailed larva of the ascidians or the appendicularians.

We might also have to reckon with the possibility that in the process of replacement of the primary larva by direct development not all adult characters were developed in the earlier phases of the life cycle. At the beginning myomery may have been such a feature, appearing relatively late in ontogeny and it might for this reason not be found in the tailed larva of the ascidians. This is merely a possibility; as far as I can see there is no evidence in this direction.

In the above I have tried to make it clear that the absence of myomery in the tailed larva, contrary to Garstang's (1928b) opinion, supplies no reason for the assumption that the chordates are the offspring of some neotenic larval form.

What has been said above is in full agreement with the older theory on the phylogeny of the tunicates. I am thus of the opinion that after the splitting-off of this group from the other chordates there appeared somewhere in their evolution a new divergence in the life cycle owing to the transition of the adult to sedentary life. The result was again two different phases, a pelagic larval and a benthic adult phase, a situation that still exists in the vast majority of the ascidians. The young in the earlier holo-pelagic life cycle became a *secondary larva* (in accordance with the definition on p. 8). Thus a new pelago-benthic life cycle resulted, but, and this must be kept in mind, a secondary one.

It is of course impossible to imagine that this transition of the adult to a sessile existence has taken place suddenly any more than it has in other sessile groups. Presumably the change was introduced by a transition to a swimming mode of life near the bottom. As has been done by Remane (1963) a parallel with certain bottom-living fishes (*Raniceps*, *Liparis*, etc.) can be drawn. A difference is, however, that the actual ancient ascidians continued right through and became permanently sessile. We must imagine that at first the sessile habit was only temporary, so that the mode of life alternated between sluggish swimming and immobility.

In such fishes as those mentioned as well as in the larvae of the ascidians and of the anurans, the viscera are concentrated at the anterior part of the body. It is possible, or even probable, that this was also a feature of the sluggishly swimming adult of the ascidian ancestors. In other words, the ascidian larvae seem to recapitulate the general body shape of these ancestors. This shape was probably adopted after the branching-off of the special line leading to the tunicates. Nothing indicates, as far as I can see, that this is inherited from the primeval chordate. (This and other problems in the phylogeny of the early chordates must, however, be more thoroughly examined than I can do here. I intend to return to the theme in another paper.)

In certain respects it is quite obvious that the present organization of the ascidian larva has arisen by alterations that took place only after the transition of the adult to a sessile existence. This applies first and foremost to the alimentary tract and the neural tube.

The alimentary tract in the recent ascidians starts to function only after metamorphosis. In the larva it consists of two portions which are not connected with each other, on the one hand the pharynx and leading from it a part bent in a dorsal direction (towards the future cloaca), and on the other the sub-chordal tail intestine (Fig. 55). The bent part is retained in the adult, while the tail intestine, which is in reality only a string of ento-derm cells, is lost at metamorphosis together with the tail.

It is possible to conclude, firstly, that originally the anus was situated at the posterior end or in any case near it, and secondly that the bent part forms a direct preparation for the conditions existing in the sessile adult. The shape of the bent part can in other words be said to be the result of the "adult pressure" that made itself felt after the adult had obtained the organization characterizing the body of the recent ascidians.

While the "bend" (as far as the shape is concerned) is a recent adult character, the tail intestine is a disappearing relic of the once functional hind part of the alimentary tract in the free-swimming ancestral form of the ascidians.

The reduction of this now evidently unnecessary part of the intestine is also an effect of that adult pressure which appeared after the transition to a sessile existence. As is so often the case with other ancient adult characters the tail intestine is not only lost at metamorphosis, but is on the way to elimination in the larva. Its reduction has probably been going on during the major part of the most recent period of evolution, i.e. practically ever since the adult became sessile.

Here it must be stressed that the phylogenetic alteration of the alimentary tract, including the shifting of the anus from the posterior end to the cloaca, took place gradually. No abrupt new formation of a secondary anus at the cloaca can have occurred. In this respect ontogeny by no means recapitulates phylogeny. The anus is a new formation only in the present ontogeny.

The part of the neural tube which is inside the tail is at an embryonic level of development. During the whole span of its existence it does not possess any nervous tissue—histologically the cells do not pass the epithelial stage—and for this reason has no nervous function. (The innervation of the lateral muscles of the tail comes from the portion behind the brain vesicle.) Thus the neural tube of the tail presents a marked parallel to the tail intestine, both histologically and functionally.

Consequently I find myself compelled to interpret conditions in the tail neural tube in the same way as in the tail intestine. In either case we have a reduction. In the recent adult the tail neural tube is missing, and it is thus quite in line with general experience to consider the reduction as the result of new (secondary) adult pressure which has arisen after the transformation of the adult into a sessile animal. I find the opposite view, which is maintained especially by Berrill (1955) quite preposterous. It maintains that in the larva of the ascidians the tail neural tube is a progressive organ which, in the evolution of the acranians and the vertebrates from this larva, as Berrill assumes, gave rise to the neural tube of the two latter groups.

Even if I were to relinquish the general principles laid down in this book I would not be able to accept Berrill's concept. Such an essential part of the central nervous system as the vertebrate neural tube cannot possibly be a phylogenetically new formation even if the ancestral form were a neotenic larva. It must be derived from a *functional* part of the nervous system in the ancestral forms.

It is equally unreasonable to suppose that the strongly altered and partially reduced intestine of the ascidian larva could have given rise to the intestine in Acrania and Vertebrata. And with this statement we can dismiss all ideas of Berrill and others about the ascidian larva as ancestral form of these groups. We can do so without entering into discussion of other organ systems and characters.

It might be pointed out that Berrill's conception is also rejected by Carter (1957). His criticism is admittedly couched in fairly general terms, but it is of great weight. I agree with it in most points without reservation.

What has been said above about the tail intestine and the neural tube should be sufficient examples of the influence of the adult pressure on special organ systems of the ascidian larva, although further examples of this kind could be given. In certain forms, however, especially within the family Molgulidae, as is so often the case also within groups with primary larvae, adult pressure has had a total effect; it has resulted in direct development mainly by the elimination of the entire tail and its contents (neural tube, notochord, tail intestine, and musculature). With this new (secondary) direct development a holobenthic life cycle has arisen. (Here I am disregarding that the eggs may have a certain floating ability.)

I shall be brief in my treatment of the two remaining groups of tunicates, Thaliacea and Appendicularia. The thaliaceans are usually regarded as holopelagic ascidians. They must, in other words, be derived from sedentary forms. (Personally I do not want to express an opinion on this question here.) It is not, however, easy to imagine that a sedentary ascidian-like form could have become pelagic by simply leaving its substratum and starting to swim. In the event that a sessile existence is in fact the primitive mode of life for the group, it appears more probable that in some form metamorphosis started to take place freely in the pelagic zone at first only exceptionally, later on more frequently, and eventually without exception. Such a change to holopelagic life is, however, possible only on condition that the adult of the benthic ancestral form was at least tolerably pre-adapted to life in the open water. Provided this was the case better adaptation could then take place easily in the ordinary way.

Irrespective of the way in which the holopelagic life cycle of the Thaliacea has arisen it is of special interest that in this group as in so many other holopelagic groups and forms the larva has been replaced by direct development (see p. 236). An exception is *Doliolum*, where a tailed larva (although somewhat altered) is still found. In other words, in this genus adult pressure has not yet taken full effect.

The appendicularians are generally supposed to be derived from neotenic larvae of ascidians or thaliaceans, and according to Garstang (1928b), who more than anybody else vindicates the thesis of neoteny in phylogenetic connections, from doliolids. Here I shall not enter more closely into this problem. I only want to say that the hypothesis of neoteny is perhaps more justifiable here than in most other cases in which it has been tried, even if the larva of *Doliolum* is hardly to be accepted as the ancestral form. This remark applies, in my opinion, also to the larvae of all recent ascidians. The reason is that the neural tube of the appendicularians is a functional organ along its whole length. It seems probable to me that the appendicularians were split off at such an early stage in evolution that the tail neural tube of the ancestral forms had not yet lost its function (cf. what has been said just now about Berrill's ideas.)

As pointed out already (see p. 200) the occurrence of cilia in the young acranian larva should be interpreted as an inheritance from the primary larva of the ancestral forms. Otherwise the Acrania exhibits considerably altered ontogeny; apart from the ciliation there is almost entirely direct development.

The young stages of *Amphioxus* are lecithotrophic and uniformly ciliated all over the body. Similar characteristics are, for instance, found in free-swimming blastulae and gastrulae of different echinoderms, but, later in the course of their development the ciliation is usually differentiated into the characteristic bands of the *dipleurula*.

In certain lecithotrophic larvae of starfishes, for example, the uniform distribution of the cilia is retained throughout the entire larval life in the same way as in the larva of *Amphioxus*. This suggests that the distribution of the cilia in somewhat older stages of this larva is no more primitive

than in such starfish larvae. In both cases it is evidently a retained embryonic feature. I have arrived above (p. 202) at the conclusion that the
primary larva in the evolutionary line leading to the Chordata was probably
of the *dipleurula* type. (A secondarily evolved retention of the uniform
embryonic ciliation is also characteristic of some other strongly altered
lecithotrophic larvae; see p. 228.)

Some words about the Vertebrata remain to be added. For this group
fully direct development is characteristic; indisputable traces of a primary
larva are not found even in the most primitive forms.*

In several instances new divergences have arisen between the juvenile
form and the adult, which have led to secondary larvae. This applies especially to the fishes and the amphibians. It is impossible to enter into a
closer examination of these groups. I shall only give a few well-known
examples in order to illustrate the fact that the conditions in the vertebrates do not differ in principle from those in the "lower" groups.

In the Urodela young and adult agree in most essential respects. Thus
a well-developed tail exists in either stage. In the evolutionary line leading
to the Anura direct development was lost, chiefly on account of the
alteration of the adult, among other things by the loss of the tail. Thus a
new divergence between young and adult came into existence, and with it
a new larva—a secondary larva.

Here one should observe the similarity with the situation in the
ascidians when their adult was altered by the transition to a sedentary
mode of life. But the parallelism does not end at this point. As in some
molgulids among the ascidians, adult pressure has in certain anurans led to
a new (secondary) direct development. At hatching it is no longer a tailed
larva that leaves the egg, but a fully-developed small frog. Here, as is the
case with many invertebrate primary larvae, the tailed larva, i.e. a
secondary larva, has become an embryonic stage. It can further be noted
that in at least one species of *Rana* from Hawaii neither external gills nor
gill slits are visible during embryogenesis. Here adult pressure has thus
taken full effect.

It may be stressed that the seondary larvae of the vertebrates are by no
means due exclusively to alterations in the adult. In many cases the juvenile

---

* An exception might perhaps be made, at least in the amphibians, for the ciliation on
the surface of the body during embryogenesis, since I consider it by no means improbable
that this is a counterpart of the ciliation which occurs in young larvae of *Amphioxus*.

forms themselves have been altered, and in other cases divergences have
resulted from alterations in both the juvenile form and the adult. It is of
course of prime importance to keep an account of the nature of the
juvenile features, since purely larval characters in the secondary larvae,
as in the primary larvae (but contrary to ancient adult characters, e.g. the
tail in the larva of the anurans) gives no information about the adults of
the ancestral forms. I shall restrict myself to certain fish larvae, where the
most marked difference from the adult consists of purely pelagic adaptive
features (Fig. 56). The greatly elongated fin filaments of these fish larvae

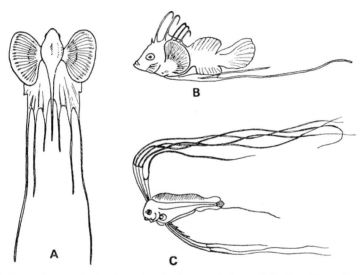

FIG. 56. Examples of pelagic adaptation in the morphology of the fry of fish. A and B,
Lophius piscatorius; C, *Trachypterus* sp. (After Ehrenbaum, from Remane, 1960.)

are undoubtedly structures that facilitate their suspension. Both in function and evolution they are analogous to the floating processes of different
kinds found in the primary larvae of many invertebrates.

There is no doubt that many larvae of fishes and amphibians exhibit
other organs and parts of organs of different kinds which, like the fin filaments in the forms illustrated, are progressive characters, adaptations
with different functions. A more detailed study of other features of this
kind would certainly be of interest.

Finally it must be pointed out that the embryonic membranes of the
amniotes are adaptive organs which in principle are of the same nature.

From this point of view the embryos of the amniotes are types of secondary larvae, adapted to a very special environment. Here, however, the life cycle has become very far removed from the original one of the metazoans.

General Section

In the introduction of this book I have presented the fundamental thesis that a life cycle with pelagic primary larva and benthic adult (pelago-benthic life cycle) is the original condition for all metazoans. An examination of the different phyla has revealed nothing to contradict this concept. On the contrary, abundant evidence points to its correctness. In the following a comprehensive summary will be given.

Emergence of the pelago-benthic life cycle and evolution of the primary larva

The pelago-benthic life cycle is spread over the whole phylogenetic tree of the metazoans. It is found in no less than seventeen phyla, which constitute the vast majority. These are Spongiaria, Cnidaria, Phoronida, Bryozoa, Brachiopoda, Platyhelminthes, Nemertini, Entoprocta, Mollusca, Sipunculida, Myzostomida, Annelida, Arthropoda, Echinodermata, Pogonophora, Enteropneusta, and Chordata. In most of these phyla this type of life cycle is more or less dominant quantitatively. In all the groups with the exception of the Chordata, in which conditions are considerably altered, an unmistakable primary larva occurs (see p. 4).

Already this wide distribution suggests that a pelago-benthic life cycle with primary larva occurred in the "primeval" metazoan, i.e. in the last ancestral form common to all metazoans. In order to visualize conditions more clearly I have constructed a phylogenetic diagram (Fig. 57) in which all groups with primary pelago-benthic life cycle have been marked with PB. (The diagram is based on facts not discussed in this book, and is open to criticism in several respects. For the following discussion the details are, however, unimportant.)

Of the sixteen groups which have been marked with PB, eight have almost exclusively a pelagic primary larva. These are Spongiaria, Phoronida,

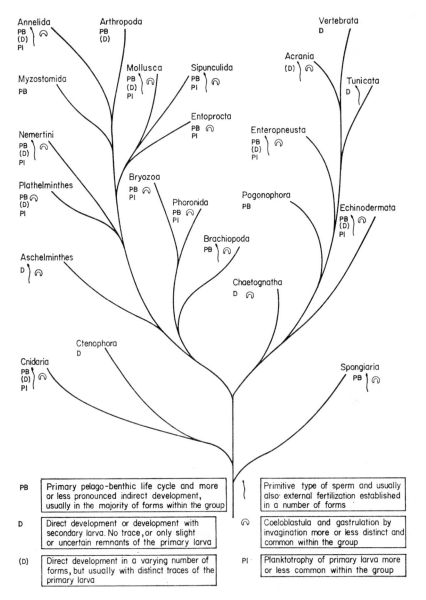

FIG. 57. Diagram of the evolutionary tree of the metazoans to show the distribution of the primary pelago-benthic life cycle and other important features. The detailed shape of the tree is not discussed in this book, but it follows, in general, the current views. Note, however, that the lines of evolution leading to Mollusca, Sipunculida, and Entoprocta have tentatively been derived from a common branch (see also p. 247).

Bryozoa, Brachiopoda, Entoprocta, Sipunculida, Myzostomida, and Pogonophora, while the same number, Cnidaria, Platyhelminthes, Nemertini, Mollusca, Annelida, Arthropoda, Echinodermata, and Enteropneusta, also exhibit numerous forms with direct development. It is, however, important to pay attention to the fact that of the latter category at least five phyla, Nemertini, Mollusca, Annelida, Arthropoda (this applies here of course to the subgroup Crustacea) and Enteropneusta, among forms with direct development contain a varying number which exhibit very distinct traces of the primary larva (*pilidium, veliger, trochophora,* etc.) characteristic of the group in question. Such traces of this pelagic larva in forms with benthic embryogenesis taking place inside the envelope of the egg, of course supply invaluable information about the life cycle at an earlier date.

Among all phyla of the metazoans there are really only three in which no trace (Ctenophora, Chaetognatha) or only a somewhat uncertain vestige (Aschelminthes) of the primary larva can be found. In the case of the Aschelminthes the uncertainty is derived from the way in which the occurrence of the adoral ciliated apparatus in certain rotifers should be interpreted (see p. 67).

The pelago-benthic life cycle is also dominant among the metazoans in the number of species. Thorson (1946, p. 431) supplies some illuminating figures for a number of groups from the region of Öresund. Among the species with a benthic adult for which the development is known the percentage with pelagic larvae is in echinoderms 89, polychaets 70, prosobranchiates 66, tectibranchiates, and ascoglossids together 92, and in lamellibranchiates 82; on the average 80 per cent of the species of these groups have pelagic larvae. Also the number of all species with benthic adult and pelagic larva must be in the neighbourhood of 80 per cent (Thorson, 1951, p. 317).

It is of great importance that in several groups direct development not only reveals traces of features from the primary larva, but that a clear

---

In conjunction with this diagram, I want to raise an objection to the fact that in a chart of the evolution of the animal kingdom prepared by Heintz (1939) and which is largely used in Sweden in elementary teaching of zoology, primary larvae and even embryonic stages have been given the rank of ancestral forms in groups and complexes. This presentation is in part based upon the concept of recapitulation in its extreme form which I am strongly refuting in this book (see, e.g. p. 248). This chart should be used with the greatest caution.

indication is supplied of the direction of the alteration. If in forms with a holobenthic life cycle the embryogenesis exhibits characters which are adaptations to pelagic life or in some way connected with it, then no doubt can remain. The secondary nature of benthic direct development is particularly clearly demonstrated by the occurrence of embryonic velar remnants in Mollusca, but also in Nemertini, Annelida, Arthropoda (Crustacea), and Enteropneusta similar remnants or other traces of the primary larva are found. (For more special facts supporting my general thesis that in all metazoans the life cycle consisted originally of a pelagic larval and a benthic adult phase, and that existing cases of holobenthic and holopelagic development are secondary phenomena, the reader is referred to the treatment of the different groups.)

It is important to note here that the *Bilaterogastraea* theory (Jägersten, 1955) forms the basis and provides a complete explanation of the conditions. According to this theory the two phases of the life cycle arose when the adult of the primeval ancestor of the metazoans, viz. the holopelagic, radially symmetrical *Blastaea*, descended to life on the bottom (and became bilateral), while *its juvenile stage remained in the pelagic zone*. Nowhere in the treatment of the different groups has this thesis met any difficulties. There are no valid reasons for the assumption that benthic direct development is an original phenomenon. An examination of the attempts to maintain such a view has proved them unfounded (see, e.g. p. 143 *et seq.*).

An account of the emergence of the primary larva and of the divergence between it and the adult in the course of evolution has already been given (see p. 4 *et seq.*), and therefore only a few fundamental points will briefly be stressed here.

Immediately after the emergence of the pelago-benthic life cycle the difference between its two phases was only slight. The ontogeny was a gradual change without any radical metamorphosis at the transition to life on the bottom. In the course of subsequent evolution both young and adult changed, however, in different ways, while still retaining the adaptation to their respective biotopes. This resulted in a gradually increasing divergence between the two stages. The young became a primary larva, and its changes were to a large extent evolutionary differentiations of special larval organs. The divergence, which can mainly be attributed to the difference between the two main biotopes of the ocean, the pelagic and the benthic zones, continued in the various evolutionary lines leading to the recent phyla. This resulted in the present primary larvae with their specific pelagic features (mainly ciliation of various kinds for swimming and feeding, and different appendages which facilitate floating). The divergence also led to metamorphosis. The greater the divergence (i.e. in most cases the greater the adaptation of the larva to the pelagic environment) the more far-reaching and the faster metamorphosis had to become.* It was in this way that evolution led to the situation now prevailing in, for example, the nemertines with *pilidium*, *Polygordius* with its large *trochophora*, and the echinoderms with the *dipleurula*. At metamorphosis the individual must not only get rid of its larval characters, but must develop at least enough benthic adult characters to maintain life. In many forms the transition to the benthic zone is facilitated by a number of adult characters that have already developed in the primary larva prior to metamorphosis.

This will be dealt with in the following section. Here we shall only recall the fact that in the case of rapid metamorphosis the individual often finds no time to utilize the larval organs and parts of the body, but must shed them in their entirety. This is the case, for example, in a number of echinoderms. In some forms an interesting mode of behaviour has arisen, consisting of the consumption of the shed parts by the metamorphosed young which in this way is provided with its first meal. Examples of this process are found in *mitraria* (see Wilson, 1932), *pilidium* (own observations; see also Cantell, 1966a), *actinotrocha*, the larva of *Polygordius*, and the *veliger*-larva of *Pterotrachea* (own observations).

Adult characters in primary larvae. The origin of direct development

Not all features of the primary larva are purely larval (pelagic) ones. In addition to these there are in many cases other features most of which are not lost at metamorphosis, but are retained in the adult. In other words, the development of some adult characters has already started during the pelagic phase of the life cycle, sometimes even in an early embryonic stage of the primary larva (e.g. shell and foot of the embryos of

---

* Here and throughout this book the term "metamorphosis" is used in the strict sense of the word. I thus apply it only to the actual transformation itself when the larva changes into the adult form, and not (as, e.g. Jeschikov, 1936) to the entire development up to this change.

many molluscs). It is important to bear in mind that this does not apply to all adult characters; the majority are only developed, as would be expected, at metamorphosis or later in the adult phase. (My distinction in this book between larval and adult characters, i.e. such which are adaptations to pelagic and benthic life respectively, of course do not exclude the existence of features in addition to these two categories. A number of transitional organs, like the larval heart and larval organs for excretion in different kinds of gastropodan larvae can perhaps not be interpreted as pelagic adaptations. Here I also want to recall the two kinds of shell glands in the larvae of *Spirorbis* which at metamorphosis form the primary tube. These glands are not an adaptation to the pelagic region, but neither can they be called adult characters in spite of the fact that they fulfil a purpose at the beginning of the benthic phase. Similar organs and others which from our point of view are without special interest have not been taken into account in this book.)

Even if the adult characters of the primary larva are only at the beginning of their development and furthermore are often modified (see p. 223), they are usually easy to identify.

We know nothing about the reason for the shifting of adult (benthic) characters to the larval phase, but it is of course possible to consider it as an advantageous change, since in this way the adult organization is prepared in good time.

This phenomenon of shifting adult features to the larva I have termed "adultation", while the unknown forces behind it are referred to as "adult pressure".

Adultation is to a large extent nothing other than a variant of the well-known phenomenon "acceleration", implying that characters which in an ancestral form appeared later in ontogeny are developed in the descendant at an earlier stage. Up till now, however, acceleration seems to have received attention only in life cycles with direct development. It is of course well known that adult characters occur also in the larvae—shell and foot of a *veliger* are for instance mentioned in almost any textbook—but this occurrence has generally been explained in another way (see below).

The implications of the term "adultation" deserve some further explanation. It is evident that for many of the adult characters that have arisen as adaptations to the benthic environment (see below) a shifting of the ontogenetic emergence to earlier stages in the pelagic phase of the life cycle has taken place (and that we thus really have a kind of acceleration).

Complications may, however, be introduced mainly by the fact that the pelagic phase in some forms is lengthened (see e.g. p. 167). In such cases there is not necessarily a question of a true shift to an earlier stage, and sometimes this is quite certainly not the case, but rather a transfer into the pelagic phase. Furthermore, in the case of holobenthic and holo-pelagic life cycles it may be difficult to decide whether or not adultation implies any acceleration. In forms with a larger reserve of yolk in the egg a general slowing-up in embryogenesis also comes into effect, and then the adultation essentially finds its expression in reduction and final elimination of the purely larval characters.

It is almost self-evident that besides *recent adult characters* a larva may also exhibit features which are no longer retained in the adult and which for this reason I have termed *ancient adult characters*. Their existence in the larva is consequently conditioned by adultation and by the well-established fact that the larva is frequently more resistant to alterations, more con-servative, than the adult. It is of great interest to find that conservatism applies also to adult characters. The importance of this fact from the point of view of phylogeny is obvious (for more details, see p. 248 *et seq.*).

In this connection it is worthy of note that ancient adult organs in the primary larvae may be assumed to have retained not only essential traits of the organization they possessed in the ancestral adults, but, at least in some cases, their original function also. A *cyphonautes* is, for instance, still able to creep about by means of the pyriform organ (see p. 42). The only difference from the ancient conditions lies in the fact that the *cyphonautes* creeps about only for a short time, until the beginning of metamorphosis, while the ancestral form continued this kind of locomotion after the transformation into the adult.

It appears as if in principle the development of all kinds of adult charac-ters could be transferred to the pelagic larva. This applies even to gonads and other sexual organs. Pertinent examples are provided by the remark-able form of Müller's larva which has been described by Heath (1928) and by some larvae of actinians (Carlgren, 1924). It is impossible to deny that the appearance of sexual maturity in larvae and other juvenile forms im-plies a shifting (acceleration) of adult characters. For this reason it is evi-dently possible in general to consider neoteny as a result of adult pressure.

According to the theory presented here the existence of adult (benthic) features in the primary larva cannot be original. In other words, their first appearance in evolution must have taken place at the beginning of the adult phase of the life cycle. There is considerable direct evidence to support the correctness of this conception, viz. all cases where the ontogenetic emergence of adult characters still takes place only after metamorphosis. As an example the tentacles in *Protodrilus* can be cited. When the larva of this genus is ready for metamorphosis it exhibits absolutely no sign of these organs. They only develop after the adoption of life on the bottom (Jägersten, 1952).*

The crown of tentacles in sabellids and serpulids provide still more illuminating evidence. In certain forms the ontogenetic emergence of this organ starts in the pelagic larva (e.g. in *Branchiomma*, see Fig. 48), while in *Pomatoceros*, for example, this happens only after the transition to the benthic zone. One of the two conditions must be the original one, and according to my opinion it is the latter one. The tentacular apparatus is without any doubt an adult character. In *Pomatoceros*, as also in *Protodrilus*, adult pressure has obviously had no effect on this organ.

Crowns of tentacles for the capture of food are in general good examples of adult characters which are first developed in the adult phase. This applies to Cnidaria (the majority of forms), Bryozoa, Testicardines among the brachiopods, Entoprocta, and many isolated species in different groups. It might also be mentioned that the so-called collar in sabellids and serpulids, which for example in *Pomatoceros* is already developed in the pelagic phase, is found in others (e.g. *Branchiomma*) only after the transition to benthic life. Attention should also be paid to the altered organization of parasites of different kinds, which is not at all or only feebly expressed in the larva (e.g. parasitic gastropods such as *Entoconcha*).

It is not difficult within almost any group of animals with a pelago-benthic life cycle to select a considerable number of adult characters of which no sign is found in the primary larva. This must in fact apply to the majority, and for this reason I consider it unnecessary to give any additional examples. If one is prepared to accept adultation, then one must also accept that the first phylogenetic emergence of the adult characters took place in the adult phase. A rejection of adultation would lead either to the assumption that they in many cases were developed in the pelagic phase from the very beginning of their phylogenetic existence (this is not only opposed by the evidence presented, but it is also unreasonable be-

---

* This applies to the majority of the species examined. *P. purpureus* which has been studied by Pierantoni (1908) is an exception. Here a change towards more direct development has taken place. Even *P. symbioticus* has an altered life cycle (Swedmark, 1954).

cause it concerns *benthic adaptive* characters) or to a contradiction of the concept that the pelago-benthic life cycle is the original one (in this case benthic adult characters in the larva would have to be explained by the assumption that holobenthic direct development was the original situation in the different groups). We should then return to the previous state of confusion over the larvae and life cycles of the metazoans.

The assumption that the first phylogenetic emergence of the adult characters took place only in the benthic phase of life is a central part in the concept of divergent evolution in the two phases of life. On the basis of my discussion the following principles can be formulated:

1. The pure larval (pelagic) characters have arisen in the pelagic larva and are with rare exceptions (e.g. Entoprocta, p. 108) not retained in the adult phase. Ontogenetically they disappear at metamorphosis.

2. The adult characters, it seems without exception, have evolved in the adult phase, but their initial development has subsequently in many cases been transferred to the pelagic larva as a result of adult pressure. Whenever an adult character has disappeared, this has usually happened first in the adult.* In this respect the larva is more conservative (note e.g. the shell of the Nudibranchiata and the alimentary tract in Pogonophora). This is the principle which conditions the occurrence of ancient adult features and which so often makes the larvae particularly valuable for considerations of phylogeny.

It might be pointed out that adultation has taken place at quite different levels of the metazoan phylogenetic tree. We must assume that in many cases parallel adultation, i.e. the independent adultation of a certain character in two or more evolutionary lines, has occurred, although it is by no means easy to prove it. An example is presumably provided by the tentacle apparatus of the Tentaculata (see p. 56). Parallel adultation must also have taken place within lower systematic units such as families, genera, and perhaps even species. For further elucidation of these problems detailed study is required. (On an interesting problem connected with the palpi of the polychaets, see p. 171.)

* It is not certain whether or not there are exceptions to this rule, but this may be so, when an increased quantity of yolk causes the loss in the larval phase of an adult character which had been shifted to this phase. Such a backward shifting (retardation) of the initial development to the adult phase seems to have taken place in the case of the tentacles of *Phoronis ovalis* (see p. 30). If these tentacles were also to be eliminated in the adult, we should have an exception from the rule.

It has been seen that in the pelago-benthic life cycle on the one hand a divergence in ontogeny has resulted in a primary larva which differs considerably from the adult, and that on the other hand adultation has produced a reduction in the difference between the two stages. In the latter case a new divergence has sometimes arisen between larva and adult, a divergence which is now, however, due to secondary alteration of the adult, most marked where there is a transition to a sessile or parasitic existence. Examples of groups which have been sessile are Bryozoa, Entoprocta, and Sipunculida. In these the adult differs considerably from that of the free-living ancestral form. The larva may on the other hand still maintain their previous appearance in certain respects. In addition to its purely larval features (and possibly occurring recent adult characters) it may still retain features which are characteristic for the adult of the free living ancestral form, such as the creeping lobe (pyriform organ) in *cyphonautes*, the foot in the larva of the entoprocts, and the head and the creeping lobe in the *pelagosphaera* of the sipunculids. In a parasitic gastropod such as *Entoconcha*, having a larva with a shell, the situation is in principle the same.

There is no doubt, however, that very often ancient adult features have also disappeared in larvae. This is the case, for example, in the strongly altered gastropod *Rhodope*, where foot and shell are entirely lost even in the larva (although a remnant of the shell gland is retained), in the larvae of certain entoprocts which also have lost the creeping organ, and in the majority of the lecithotrophic bryozoan larvae where the *cyphonautes*-shells have disappeared. We must imagine that in these altered forms adult pressure has been changed, or rather is operating in a new direction. While at one time adult pressure had the effect of shifting the development of a particular adult character to the larva, it has, when the adult character in question is no longer found in the adult, caused its reduction or even elimination. The important conclusion of this is that, *once an adult character has been lost in the adult it cannot appear in the larva.*

Primary larvae may have undergone adultation in one or several features. Where several are involved, the larva reaches such a high degree of similarity with the adult that it becomes questionable whether development should be termed direct or indirect. Whether or not a larva can retain a pelagic mode of life after strong adultation depends partly on the original nature of the adult features and partly on their potentiality for modification; it may also depend on whether or not they can be compensated by other features. *Lingula* is a form where the larva is still pelagic in spite of its numerous adult characters (tentacular apparatus, mantle, shell, setae, peduncle, etc.—see p. 47).

The fact that in the larvae the adult characters can be modified seems to be a common phenomenon. On further consideration it seems quite understandable that they are altered in such a way as to fit a pelagic existence or at least not to constitute any direct obstacle. The modification as such is of course a newly-acquired larval character, an adaptation to the pelagic zone. In certain cases the modification can become so great that the original adult character (organ, part of body) cannot be taken over directly by the benthic adult, but is lost at metamorphosis, only to be regenerated afterwards, unless it had already been regenerated beforehand. An example of such a "pre-regeneration" is given by the tentacles of most *actinotrocha*-larvae (see p. 27).

It was mentioned that *Lingula* is an example of a form in which several adult characters are prepared during the pelagic phase. It might be considered natural for this situation to lead to the total elimination of the pelagic phase, so that the embryo attached itself after the hatching from the egg. This might result in holobenthic development, perhaps after some more adult characters had been transferred to the larva. It is uncertain whether or not this kind of phylogenetic "weighing-down" of the larva to life on the bottom by an accumulation of adult characters (and by prior or simultaneous elimination of the pelagic characters of the primary larva) has taken place anywhere as a means of evolution to benthic direct development. Nothing, however, seems to point in this direction. (Change to holobenthic life must have more frequently taken place by enclosure, see p. 227). It seems, on the contrary, that in *Lingula* and its relatives, in spite of the numerous benthic adult characters in the larva, a certain lengthening of the pelagic phase has occurred.

I have here entered into a by no means unusual phenomenon. A lengthened pelagic phase (and with it considerable growth) coupled with accumulation of adult characters in such a way that the original larval nature is lost to a considerable degree is found, for example, in many annelids: spionids, *Magelona*, *Poecilochaetus*, etc. *Poecilochaetus* is a very extreme example, in that the transition to life on the bottom is postponed until the specimen has reached a length of about 10 mm, and a large number of segments have developed. In, for example, certain spionids even sexual products may be formed in the course of the lengthened pelagic phase (see p. 167). There is, however, no certain proof that this lengthened pelagic existence in larvae has led to holo-pelagic life cycles (cf., however, p. 234).

The above discussions of the different groups has shown that there is an obvious correlation on the one hand between paucity of yolk in the egg and planktotrophy, and on the other between an abundance of yolk and lecithotrophy (see also p. 270 *et seq.*). We have, furthermore, found support for the assumption that planktotrophy is the original condition in at least the following groups: Cnidaria, Phoronida, Bryozoa, Brachiopoda, Plathelminthes, Nemertini, Entoprocta, Sipunculida, Mollusca, Annelida, Echinodermata, and Enteropneusta (i.e. in the majority of phyla with primary larva).

In terms of the number of species planktotrophy is strongly dominant; according to Thorson (1951) planktotrophic larvae occur in over 70 per cent of all benthic invertebrates.

All this indicates strongly that the ancestral form common to all the groups listed possessed a planktotrophic primary larva. This may be con-sidered as implying that the generalization holds good for all groups of the metazoans with the probable exception of the Spongiaria (see phylo-genetic tree, Fig. 57). On p. 12 I have discussed why an exception should probably be made for the sponges. The alimentary tract, an ex-tremely old adult character, must only have been transferred to the larva after the splitting-off of the sponges. The first pelago-benthic metazoans (the *Bilateroblastaea* forms) were without an alimentary tract throughout their entire life cycle. At this stage of evolution nutrition must have been supplied by phagocytosis in the cells of the surface of the body, perhaps supplemented by autotrophy (see Jägersten, 1955). When the alimentary tract (and with it the mouth) evolved at the emergence of the *Bilatero-gastraea* forms, it was restricted to the benthic phase of the life cycle. (At this evolutionary stage and immediately after, gastrulation formed the essential part of the metamorphosis which took place at the transition to the bottom.) "Within a short time" adult pressure resulted in the ac-quisition of the alimentary tract by the pelagic larva.

Once the larva had obtained the alimentary tract, the mode of feeding could be changed to uptake of particles through the mouth, i.e. plankto-trophy appeared. The originally quite small amount of yolk in the egg (see p. 241) necessitated the uptake of food at a relatively early stage of ontogeny.

The original planktotrophic mode of feeding was then altered in different ways into more or less complicated mechanisms in the various lines of evolution. It was these mechanisms which, together with organs for swimming and floating, accounted for a large part of the divergence between primary larva and benthic adult (see above, p. 216). It is perhaps essentially the "merit" of planktotrophy that the pelagic phase is still retained in so many instances.

In a great many cases, both in whole groups and in isolated species, planktotrophy has, however, been replaced by lecithotrophy. For this change it was necessary for the quantity of yolk in the egg to have increased sufficiently to ensure adequate nutriment up to the termination of metamorphosis, when food could be taken up in the benthic biotope.

A larva of this kind no longer had to take up food from outside. For this reason the apparatus for capture of food, no longer being required, could be gradually reduced without disadvantageous results. This reduction lessened the difference between larva and adult, and thus constituted a first step towards direct development.

Numerous forms, distributed over several phyla, still occupy an evolutionary stage of this type with lecithotrophy in a free-swimming, though simplified larva. The change to lecithotrophy, etc. has occurred independently in many different instances. Often the change has taken place at such a late evolutionary date that of two closely-related species one is still planktotrophic, while the other is lecithotrophic. This applies, for example, to species of *Protodrilus*.

In connection with the increase in the amount of yolk and change to lecithotrophy the larva has in many cases lost the lumen of the alimentary tract and frequently also the mouth (e.g. *Octocorallia* and many bryozoans). Their ontogenetic emergence has, in other words, returned to the adult phase. Abundance of yolk and lecithotrophy seems able also to effect a retardation in development of other organs (see p. 30).

As far as is known the change to lecithotrophy is an irreversible process. At all events, I have not been able to find any evidence of a return of planktotrophy in the *original* larval phase. On the other hand it cannot be excluded that in isolated cases planktotrophy may have reappeared in forms which have acquired a lengthened pelagic phase. Widersten (1968) considers this to be the case in some actinians.

There is no reason why larvae which have changed to lecithotrophy should be pelagic for the sake of feeding. The fact that many still do so evidently depends on the slowness of the changes, in certain cases perhaps on the necessity of spreading (see below). On the other hand it must be noted that a number of forms have ceased to be pelagic, having become holobenthic. There are here all transitions both in mode of life and reduction of the pelagic features. Thus several more or less altered larvae are known which for a short time live a "bottom-pelagic" life. The next evolutionary step is holobenthic direct development. (For direct development in forms with pelagic adult, see p. 236.)

Thorson writes that the majority of the lecithotrophic larvae pass "eine lange Zeit in plankton" (1951, p. 285). I should like to put a query after this statement. There are certainly a great number of larvae of this type which live pelagically only for a very short time, often only for one or two hours. I need only remind the reader of the numerous lecithotrophic bryozoan larvae and of larvae like those of *Spirorbis*, *Echiurus*, *Antedon*, and *Saccoglossus*.

In certain species the life cycle has been found to be pelago-benthic in one part of the population and holobenthic in another, and this within one and the same area (see pp. 126 and 173). Such species are evidently in a transitional evolutionary stage.

We do not know much about the cause behind the increase of the yolk in the egg which has often taken place. It must be admitted that it often seems to be connected with the environmental conditions, but such a connection is by no means always the case. Generally the quantity of yolk increases with immigration into brackish and fresh water and in arctic and antarctic regions, but numerous purely marine forms with large yolky eggs are also known from temperate and warm waters.

On the other hand the amount of yolk has in certain cases diminished. This secondary paucity of yolk occurs in forms where the mother animal supplies the embryo or larva with special nutriment, such as abortive eggs or albumen secretion. It occurs also in viviparous forms with placental structures (phylactolaematous bryozoans, vertebrates). It is interesting to find that remains of features of the pelagic larva (e.g. of the velum in the embryo of *Paludina*) are sometimes encountered even in ontogenies which have been greatly altered by vivipary.

It is hardly probable that the quantity of yolk as such exerts any real influence on adultation. Perhaps an ample supply of yolk provides adult pressure with a wider scope. Adult pressure might be said to meet with less resistance in the reduction of the pelagic larval characters. In spite of this, *total adultation* has not occurred so very often, if by this we mean that a direct development no longer contains any trace of the primary larva.

Among the invertebrate groups the *Chaetognatha* supply the best example of total adultation in all species. Within many groups this kind of change has, however, happened in isolated species (see, e.g. Fig. 51, p. 191).

The appearance of lecithotrophy in the larva is as a rule connected with a considerable shortening of the pelagic phase, or even with its total elimination, at least in free-living forms (cf. below, on dispersal). A lecithotrophic larva can be enclosed in egg membranes and cocoons of different kinds or in the interior of the mother animal without any detrimental result, and may in this way become an embryo. Such enclosure has happened in a large number of cases, both in isolated species and in whole groups of different taxonomic rank. Complete enclosure results in the hatching of a juvenile which is perfectly ready for benthic life. In such cases it is customary to speak about direct development, even if remains of pelagic larval characters still appear during the enclosed phase of the life cycle.

In the foregoing I have carefully avoided saying that enclosure and the increased amount of yolk in the egg *give rise* to the reduction of the pelagic organs and features. We really know nothing about the causal connections, only that enclosure is usually *accompanied* by a reduction of this kind. (A similar idea is expressed by Abeloos, 1956, p. 179.)

It might be stressed that enclosure may also take place in forms with planktotrophic larvae, but here only part of the development can become enclosed. The possible duration of the enclosure depends of course on the amount of yolk and possibly of other food supply deposited inside the egg envelopes by the mother animal. Larvae which are, so to speak, on their way towards acquiring lecithotrophy may of course be enclosed during the greater part of larval development.

I have just mentioned that enclosure is in general followed by reduction of the larval (pelagic) characters. There are, however, cases in which even the organs which had originally functioned in planktotrophy are not only retained after enclosure, but also continue to function. In the foregoing (p. 127) I have alluded to a case where a *veliger* still makes use of the velar apparatus for feeding during its life inside the cocoon. In this case it must have been a planktotrophic larva that was enclosed.

A closer analysis of enclosure and its evolution would certainly be of interest, but would be too great a digression here. I only want to stress that enclosure, no matter of what kind, cannot be considered as primitive. The original situation is doubtless a life cycle without enclosure with the eggs (more or less poor in yolk) being discharged freely into the water (see below). Where they are surrounded by a membrane, as is probably always the case after fertilization, this membrane is often pierced by the cilia of the embryo. This accounts for swimming at a very early stage Examples of such conditions occur, for example, in the cnidarians, annelids, and sipunculids. The next visible step in the series of phylo-genetic alterations is that the locomotory cilia no longer penetrate through the egg membrane and for this reason at first merely cause the embryo to rotate inside the membrane. If in such cases early hatching takes place, this does not constitute, biologically, any essential deviation from the original condition. Conditions of this kind occur within quite a number of phyla, e.g. Phoronida, Bryozoa, Nemertini, Mollusca, and Echinodermata.

It must be fairly obvious that eggs which are discharged freely into the water represent the original situation (see also p. 240 *et seq.*). It is less easy to say whether or not forms which at one time had acquired some kind of envelope around the eggs could have reverted to the original state. As far as I can see, nothing indicates this. Holopelagic forms such as Janthinidae and the heteropods, speak against this possibility. One might think that a return to free eggs should have happened first of all in such forms, but there is no supporting evidence. The heteropods lay their eggs enclosed by strands of secretion, and among the Janthinidae some of the species are even viviparous.

While, as mentioned above (p. 223), it is uncertain whether or not benthic direct development has ever arisen through the accumulation of adult characters in *planktotrophic* larvae, it is quite certain that this has happened many times via *lecithotrophic* larvae which in the way described have become entirely enclosed.

There are, however, other courses of evolution which have resulted in or are leading towards direct development via a lecithotrophic larva, though total enclosure does not occur. In certain echinoderms, eggs more or less rich in yolk give rise to larvae which are considerably simplified in their external appearance (see p. 190). They are ciliated all over the body and after swimming about for a time they are changed into the adult form without any true metamorphosis. A similar sequence is found in *Amphi-oxus*, in which the development is still more direct. In either case the larva must be derived from the *dipleurula* type, and the uniformly distri-buted ciliation seems to be a retained embryonic feature.

Similar larval forms (lecithotrophic and with ciliation all over the whole or the greater part of the body) are found in quite a number of sessile species and groups. This type includes the aberrant larva of *Phoronis ovalis* (p. 28) and certain lecithotrophic larvae of bryozoans, *inter alia* those of Phylactolaemata. They differ very much from the planktotrophic larvae of related forms (*actinotrocha* and *cyphonautes*, respectively). It is hardly possible to consider the divergences (simplified shape and uniformly distributed ciliation) as conditioned by adult pressure, and for this reason it is hardly a question of evolution towards typical direct development. In the case of a group like the Bryozoa it is in fact hardly possible to think of fully-accomplished direct development. Here, if anywhere, a free-swimming stage is necessary for dispersal (cf. p. 232). The fact that in Phylactolaemata polypids are already developed in the larva proves, however, that adult pressure exists and in special cases can also have an effect on pronouncedly sedentary forms.

By way of conclusion to this section about alterations of the pelago-benthic life cycle the following speculation might be permitted. If the frequency of the phenomenon of adultation and of its strongest manifestation, direct development, is borne in mind, it is difficult to escape the impression that strong "antilarval" forces lie behind what has been termed "adult pressure". These forces must be considered as having been active ever since the metazoans arose. The fact that pelagic primary larvae are nevertheless retained in so many cases indicates that conservative "prolarval" forces have also existed and presumably still exist. One might say, metaphorically, that there is a combat between these two different forces. In this combat the antilarval forces have been victorious in many cases.

Secondary larvae. Secondary direct development

In the majority of groups and species there has been no further radical alteration of the life cycle after the appearance of direct development. In certain cases, however, the alteration has continued and has resulted in a new divergence between juvenile form and adult to such an extent that the former has again assumed the character of a larva. A larva of this kind I have called secondary larva. As would be expected, the difference between the two stages has become particularly marked in the cases where they have become adapted to different biotopes and/or have assumed essentially different modes of life. At the emergence of secondary larvae adult pressure has had to give way to other forces.

In principle a secondary larva may arise by an alteration either of the juvenile phase or of the later part of ontogeny (the adult phase) or both. The result was in any case a new metamorphosis, a secondary metamorphosis, provided the divergence had gone "far enough". What had happened was to a certain extent a repetition of the evolution which had previously led to the life cycle with a primary larva.

It seems as if most of the secondary larvae arose first of all as the result of alteration of the adult. Such cases are of great phylogenetic interest, since then the larva is the carrier of a varying number of adult characters dating back to the earlier phase of evolution with direct development, before the adult was altered.

Perhaps the best example of such a secondary larva is found in the Ascidiacea. Here its formation can be attributed to the fact that the stages in the later part of the life cycle have become sedentary, with a very great transformation of the adult as the result. The juvenile form, on the other hand, has retained the pelagic mode of life and the essential features of the organization which existed during the evolutionary period with direct development (see p. 206). This implies that the pelago-benthic life cycle of the ascidians is secondary.

Other examples of the emergence of secondary larvae by alteration of the adult are found in amphibians (Anura), certain ctenophores, and rotifers (see pp. 23 and 66, respectively), and probably also in Priapulida (see p. 72). Some parasitic forms within groups with direct development have acquired larvae of this kind, for example, the epicarideans within the Isopoda. The epicarideans are, however, of minor interest from the point of view of phylogeny, since the larvae hardly provide any information about the ancestral forms beyond what is known already from other sources. The larvae of the Nematomorpha, on the other hand, can supply a certain amount of such information (p. 75 *et seq.*).

It is perhaps less common to find that in a direct development the juveniles have become altered to such an extent that they merit the name of secondary larvae. Some examples can nevertheless be given: the cercariae of the trematodes (p. 82), the *phyllosoma*-larva of the loricate decapods (p. 178), and the larva of certain fishes (p. 211). In these cases at least there are progressive adaptive characters, appliances for swimming or floating, features parallel to similar structures in the primary larvae.

It should be noticed that no reappearance of the locomotory ciliation seems to be possible in the secondary larva once it has disappeared. Where the secondary larva is pelagic, swimming must therefore be effected by some means other than cilia (see, e.g. the cercariae).

It can be seen that it is possible in principle to distinguish between purely larval characters (new progressive characters, usually of an adaptive nature) and adult characters both in secondary and in primary larvae. The adult characters may be either recent or ancient, the ancient ones being of greater interest.

It is uncertain whether, among the ancient adult characters of a secondary larva there might not also be hidden features derived from the pelago-benthic period which preceded that with direct development. I have at all events only found features from this latter period.

Attention might be drawn to the fact that the new divergence between young and adult which led to the secondary larva also resulted in a new adult pressure. This "aims" at the elimination of the secondary larva in exactly the same way as the adult pressure which previously caused the disappearance of the primary larva. This has succeeded in certain cases, where thus new direct development (secondary direct development) has been brought about. This applies, for example, to certain ascidians and anurans (see pp. 208 and 210, respectively).

In practice, the classification of the larvae sometimes meets with difficulties, even though it may be quite easy to define what is meant by the term "secondary larva". This is due to the fact that here as everywhere, where evolution is concerned, transitions can exist. Furthermore, since it may be doubtful whether or not a life cycle with strongly altered larva previously contained direct development, there may be some hesitation in the interpretation of this larva. An example is provided by the *glochidium*-larva of the unionids (p. 129). Such cases are, however, rare.

To summarize, the secondary larvae can be of two different types:

1. Those which in one or more respects recapitulate an ancient adult (they have come into existence owing to an alteration of the adult in a life cycle with direct development).
2. Those which do not recapitulate an ancient adult (they have arisen through the transformation of the juvenile form in a life cycle with direct development).

The two categories are, however, not sharply delimited. Intermediate forms are, at least, conceivable.

The pelagic larvae and dispersal

The opponents to Haeckel's ideas about recapitulation have often pointed out that the larvae must be considered as pelagic adaptations for spreading the species. De Beer (1958, p. 41) writes, for example: "These larval forms" (he refers here to *trochophora* and *nauplius*) "as well as many others such as the pilidium, veliger, pluteus, actinotrocha, must be regarded as larval adaptations to dispersal, and not, as recapitulationists contend, as the representatives of ancestral *adult* forms."

In this book I have many times stressed that the larvae are adapted to pelagic life, but this is not the same as saying that they are "adaptations to dispersal". According to prevailing opinion pelagic larval life should have a selective value, which in several different instances has caused the transformation of benthic juveniles in the existing pelagic larvae.

It is of course perfectly true to say that a pelagic larval life is of importance for dispersal, but the question is, whether or not it really has a selective value, i.e. is more favourable than a benthic larval life. If there is a selective value, it must often be obstructed in some way, since it is known that in many cases evolution has progressed *from* development with a pelagic larva *to* a holobenthic life cycle. There is, on the other hand, no indication that primary larvae have originated from juvenile forms in cycles with benthic direct development. For this reason it appears that a holobenthic life cycle would be rather more advantageous—if we can speak about advantages and disadvantages, which I am unwilling to do in connection with studies in phylogeny. And if my view, that for all metazoan groups the division of the life cycle into a pelagic larval phase and a benthic adult phase is the original condition, is valid—and everything points in this direction—*the need for dispersal is already satisfied at the outset.* The situation is: if advantages are attached to a pelagic mode of life, then these have contributed to the retention up to the present of the primary larva in the majority of the marine groups. The observation that it has nevertheless been replaced in so many cases by benthic direct development might lead to the suspicion that possible advantages connected with the latter type have often prevailed. The "antilarval" forces might be said to have predominated.

Evidently the dispersal necessary for survival is also possible in forms with holobenthic development. Already a very slow locomotion seems to suffice. In sedentary forms the need for dispersal can be fulfilled by the floating eggs.

Häcker (1896) took a critical view of this problem. He pointed out that the tubicole polychaets, which in the adult stage are at best only slightly mobile, and for this reason would have a particular requirement for pelagic larvae, in fact often are without them. This Häcker sees as an indication that dispersal may not be the only advantage, perhaps not even a very important one, of the pelagic mode of life. Because larvae that are hatched from large, yolky eggs develop slowly and swim about only for a short time or not at all, while larvae from small eggs, poor in yolk, develop quickly and have a long pelagic existence, Häcker concludes that the larvae are driven into the upper layers of the water by the need for food. Häcker is evidently of the opinion that only larvae from small eggs need to swarm about, but not those from large eggs which thus can remain upon the bottom.

Häcker is partly right. There is no doubt that nutrition is the decisive factor, but it is not correct to say that larvae hatched from small eggs, poor in yolk, have assumed a pelagic life. Pelagic life is the original situation everywhere. At a later stage of evolution an ample supply of yolk has in many cases allowed the transition to a holobenthic life cycle.

The above applies to primary larvae. In secondary larvae the situation seems to be different, at least in some cases. It is thus possible that the cercariae of the digenetic trematodes, or rather their swimming tail, have evolved for the "purpose of dispersal". I consider it less probable, however, that the ciliation of larval forms like *miracidium* and *coracidium* is a new feature for the same purpose. It has been mentioned (p. 82) that it might be a feature retained from the original pelago-benthic life cycle (in the *oncosphaera* of the taeniids it has been lost).

The holopelagic life cycle

In the metazoans the pelago-benthic life cycle has been changed not only into a holobenthic, but in several cases into a holopelagic life cycle. This has happened within evolutionary lines which have led to systematic units of very different rank. But among the phyla only two—Ctenophora and Chaetognatha—have been altered in this way. The occurrence of benthic forms within these phyla, Ctenoplana, Coeloplana, Tjalfiella, and Spadella, respectively, depends on a return to life on the bottom.

The adoption of a holopelagic life cycle might be imagined to have taken place in different ways. Often the opinion is expressed that this occurred via neoteny, i.e. that a pelagic larva became sexually mature and then failed to develop into a benthic adult. Above I have discussed this possibility and found that a transformation of this kind *could* have taken place, for example, in *Arachnactis* within the Cnidaria (p. 14) and in the evolutionary line leading to the Ctenophora (p. 24). In these cases the conditions would have arisen directly from a pelago-benthic life cycle with a primary larva. If *Graffizoon*, described by Heath (1928), which is exactly like a Müller's larva except that it possesses sexual cells, does not

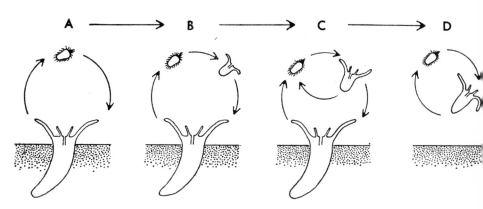

Fɪɢ. 58. Diagram showing the probable course of evolution of a holopelagic life cycle from the original pelago-benthic life cycle through elimination of the benthic adult via neoteny (see also text).

develop into a benthic adult (there is here the same uncertainty as in the case of *Arachnactis*, see below), this would present an example of the origin of a holopelagic life cycle via neoteny in a primary larva.

Figure 58 shows how this course of evolution might proceed. In A we have the original, uncomplicated pelago-benthic life cycle, in B a lengthening of the pelagic phase has taken place, and certain adult features, e.g. the tentacles which in A were developed only after the transition to life on the bottom (cf. p. 221) now start to develop during the free-swimming phase. Stage C is characterized by sexual maturity of the large tentaculate larva which, however, in part of the population is still transformed into a benthic adult. At stage D this no longer happens, and the life cycle has become holopelagic. It must be imagined that in the period between C and D a decreasing number of individuals descended to the bottom.

This diagram represents most closely the development of *Arachnactis albida* (assuming that this form really is holopelagic), but may be applicable also to Ctenophora. I must stress, however, that we have no information about the nature of the original benthic adult in this group (cf. p. 24 above).

If the tunicate group Appendicularia has arisen from a neotenic larva belonging to some sedentary ancestral form (see p. 209), we should have another example of the origin of a holopelagic life cycle from a pelago-benthic one, but, and this must be borne in mind, via a secondary larva.

In the case of the pteropods and some other gastropods I have come to the conclusion that the most likely course to their holopelagic mode of life was a gradual alteration of locomotion and morphology in the *adult* of a primary pelago-benthic life cycle. I refer the reader to earlier discussions (see p. 130). A gradual transition to holopelagic life is of course possible only in forms with a free-living adult.

In theory it might be imagined that a pelago-benthic life cycle gave rise to a holopelagic one in such a way that the individuals after metamorphosis in the open water remained there for the rest of their lives. Such an alteration should be gradual in that initially only a minority of individuals no longer assume life on the bottom, but their number increases with time until eventually the entire population remains in the pelagic zone. Evolution of this kind demands that the adult which is adapted for a benthic mode of life already possesses features that will enable it to assume a pelagic existence without immediate alteration. In other words, it must in essential respects be pre-adapted to pelagic life. In this case there is no neoteny, even if in other respects there is some agreement with the process described in Fig. 58.

I do not consider it impossible that for instance, Siphonophora, and Thaliacea, and perhaps also the hydroid genus *Pelagohydra* might have undergone evolution in this way, even if I cannot find any evidence for it. (Thaliacea: see discussion on p. 208.)

A closer discussion of the way in which the various holopelagic metazoans might have come into existence is outside the scope of this book. In the present incomplete state of our knowledge this is indeed hardly possible. The foregoing is only intended to show that in principle an origin via neoteny as well as gradual ascension of the benthic adult both appear possible. However, I consider that origin via neoteny has frequently been accepted in the literature without sufficient reason. The pteropods might provide an example of this.

However the holopelagic life cycle may have come into existence, it is certainly secondary.

It is remarkable that in holopelagic life cycles the development is very often direct. (I have referred to this point earlier in the treatment of some different groups; see p. 24 and 180.) With regard to the cause, literature often expresses the opinion that a larva is no longer "necessary", since in life cycles of this kind dispersal is effected by the adult.

I do not find this explanation satisfactory. My view of the problem is as follows. The situation varies according to the way the holopelagic life has evolved. If it has happened via neoteny in a pelagic larva, direct development automatically results, metamorphosis and the benthic adult being simply eliminated. It is, in other words, the larva which becomes the new, secondary adult.

The situation is quite different when there is a gradual transition to pelagic life by the adult in a pelago-benthic life cycle. In such cases the difference between larva and adult and consequently also metamorphosis is initially retained. (This is to a high degree still the case in pelagic gastropods; see p. 130 *et seq.*) With the transition to the pelagic mode of life the adult becomes altered in response to the demands of the new biotope, especially in its locomotion and feeding, and gradually becomes well adapted to the pelagic region, *although in a different way from the larva*. The ever active adult pressure has the effect of shifting the development of the new (pelagic) characters of the adult to increasingly younger stages, where they fit just as well as the larval features. For this reason adult pressure, so to speak, meets less resistance than in the numerous cases where benthic adult characters are to be transferred to a pelagic larva. Since it is unnecessary, perhaps even unsuitable that larval and adult pelagic characters exist side by side in a holopelagic life cycle, gradual reduction and eventual elimination of the larval characters takes place. In other words, ontogeny leads directly to the adult stage.

It seems to be the rule that pelagic adults have larvae or young which are likewise pelagic. The most probable explanation is that holopelagic life has always arisen *directly* from pelago-benthic life cycles. As far as I can see there are no indications that evolution via holobenthic life has ever taken place.

Criticism of earlier opinions about the life cycle

In the foregoing it has been pointed out that the evolutionary process called here adultation (i.e. acceleration of development of adult characters to the pelagic larva) has received no attention until the present. The reason for this seems to be that it was taken for granted that benthic direct development was the primitive condition. This was perhaps a more or less unconscious reaction against the extreme recapitulationists. In any case it has often been believed that the primary larvae must be juvenile forms which had ascended into the pelagic zone and, in so doing, taken with them some previously arisen benthic adult characters. This view is held, for example, by Hadži (see below). The special characters which are adaptations to pelagic life (ciliary rings, etc.) would have arisen during and after the ascension. Schultz (1910) even says quite positively that the acceleration would not be able to influence the larvae. The increasingly earlier development "trifft aber oft wieder auf eine Schwelle beim Larven-stadium, auf welches, bei den besonderen Lebensbedingungen desselden, sie nicht übertragen werden können ". This assertion is in my opinion quite wrong.

I have not come across any other opinions about this important question from the authors who discuss acceleration. In his section on this phenomenon de Beer (1958), for instance, says nothing about the pelagic larvae. Neither does Remane allude to the conditions in these larvae in his paper about the biogenetic rule (1960).

On the subject of direct development, it might be pointed out that it had been proposed earlier, in a different context that this type of development is a secondary phenomenon resulting from the elimination of the pelagic larva. Several authors well realized that, for example, the occurrence of a velum in the embryos of gastropods with a holobenthic life cycle proves the previous existence of a free *veliger*-larva. The only new element in this part of my general concept is the thesis that the pelago-benthic life cycle must be considered as original for *all* metazoans.

The reason why a consistent theory of the life cycle and of its evolution in the metazoans has not been presented earlier, is in my opinion that the earlier investigators who dealt with this problem did not have a sound understanding of the early phylogeny of the metazoans. It was only by an idea presented in connection with the *Bilaterogastraea* theory that a basis for a comprehensive scheme was supplied (Jägersten, 1955, 1959).

Haeckel and his adherents were hampered first and foremost by their extreme idea of the recapitulation. According to this the present pelagic larvae, provided they had not been secondarily altered ("caenogenetic"), should recapitulate the respective ancestral forms in the adult stage. Thus these ancestors should have been holopelagic. The common ancestral form for all *trochophora* groups, for instance, should have resembled a *trochophora* in all essential features except that it was fertile. This imaginary ancestor which has even been given a special name, "*Trochozoon*", has played a prominent rôle in the discussion.

This concept was connected with other improbable speculations, such as the corm theory of the annelids. The segmented body should thus be a chain-like colony of more or less reduced individuals of *Trochozoon*.

Haeckel's concept about pelagic larvae was accepted by many, but rejected by others. The opponents maintained in general that the shape of these larvae can be attributed exclusively to adaptation to their special environment and is without phylogenetic importance. In later years Haeckel was attacked especially by Hadži (1963 and earlier papers) and de Beer (1958). They oppose not only the doctrine of recapitulation but also the *Gastraea* theory. The opposition is, however, based on an untenable starting point, viz. Hadži's idea of the origin of the metazoans from protozoans of ciliate type via acoelous turbellarians.

According to this ciliate theory, which is strenuously defended by Steinböck, holobenthic direct development must be an original feature in the metazoans. The planuloid theory (Hyman and others) leads to the same conclusion in the case of the coelomates, since according to this theory these would also have arisen via holobenthic acoelous forms. The demonstration that a pelago-benthic life cycle is probably the original condition for all metazoan phyla is a further proof of the unreliability of both the ciliate theory and the planuloid theory, but supports the *Bilaterogastraea* theory.

Scientists who have examined only some special group or form and who have not taken the problem of the evolution of the metazoans into consideration, quite often advance the idea that holobenthic development was the original condition and that the pelagic larvae have arisen in parallel in different places. They even go so far as to propose parallel evolution for such a uniform type as the *trochophora*. (In the foregoing I have had the opportunity to oppose such ideas, see pp. 140 and 143.)

According to the current of opinion the juvenile form has at different stages within the metazoan evolution changed from pelagic larva to benthic young and vice versa. I do not consider that this is correct; the existence of the primary larvae must be explained in another way, as has already been discussed in detail. It is, however, not incompatible with my general concept that the juvenile in a benthic direct development should again be able to become a pelagic larva. In such a case it would, however, be a secondary larva, and it is not easy to find reliable examples of such a course of evolution. Perhaps one could be found in the cercariae of the trematodes (see p. 82).

Finally, it is worthy of comment that in such groups as Oligochaeta and Hirudinea, which have lost the pelagic phase of the life cycle (presumably in connection with their migration into fresh water), there is not a single form which has regained a pelagic larva—not even in cases where a return to marine life has occurred.

The connection between certain ontogenetic phenomena

In the treatment of the different groups I have found it important to draw attention to the occurrence of certain features in the early phase of the life cycle which are all more or less interconnected. We are concerned with a number of phenomena of such importance from the point of view of phylogeny that we must make an attempt to elucidate what is original and what is altered. In the following table I have listed and classified these phenomena according to the results of my investigations. In this way two parallel series emerge. The features in the A-series occur in the original type of life cycle, i.e. the pelago-benthic cycle with primary larva, while the points in B-series occur mainly in cycles with benthic direct development or with secondary larva. (N.B. any given ontogeny need not conform in all respects to one of the series.)

The phenomena in the A-series together constitute a system (the "primary system" of the ontogeny) in which for certain only one point (external fertilization) can be altered without an inevitable change occurring in one or several other points. In the following each phenomenon will be examined in this respect and also as to whether or not it is original.

1. Enclosure of the egg and a greater or lesser part of the ontogeny in envelopes of different kinds (secondary egg membranes, cocoons, etc.) or

A. ORIGINAL CONDITIONS (Primary system)	B. ALTERED CONDITIONS (Secondary system)
1. Eggs freely discharged into the water and free pelagic development.	Eggs not freely discharged. Development occurs for a varying length of time in envelopes of different kinds or inside the mother animal ("enclosure").
2. External fertilization.	Internal fertilization of more or less complicated type.
3. Small quantity of yolk in the egg.	More or less ample quantity of yolk (at enclosure perhaps supply of other nutriment to the embryo).
4. Total and equal cleavage of the egg.	Unequal or partial cleavage.
5. Blastula with spacious blastocoel. Cilia on the surface.	Blastocoel more or less narrowed or even absent. No cilia.
6. Gastrulation by invagination.	Gastrulation of other types.
7. Planktotrophy in the larva.	Lecithotrophy in the larva or more or less accomplished direct development.

in the mother animal entails complications in the mode of fertilization, generally a change to internal fertilization, and with it also a transformation of the original type of sperm (Franzén, 1956). Enclosure presupposes an augmentation of the yolk in the egg in proportion to the duration of enclosure. (Later in the course of evolution the quantity of yolk may again diminish if the mother animal supplies the embryo with nutriment in some other way.)

The discharge of free uncovered eggs is without doubt the original situation in the metazoans. (Probably the eggs of their primeval ancestor were floating.) As far as I can see there are no indications that once enclosed eggs have ever become secondarily free.

Free eggs are still retained in forms of at least the following groups: Cnidaria, Ctenophora, Phoronida, Bryozoa, Brachiopoda, Nemertini, Mollusca, Sipunculida, Myzostomida, Annelida, Echinodermata, and

Chordata. Within the majority of these groups enclosure and brooding has been introduced in a varying number of forms, and has been elaborated in a more or less complicated way.

Another less pronounced deviation from the original condition is the deposition of the eggs on the bottom without enclosure (e.g. certain species of *Protodrilus*).

2. Although the mode of fertilization is changed in connection with enclosure, internal fertilization may on the other hand arise, so to speak, spontaneously. An example of this is provided by *Myzostomum*, where an extremely complicated mechanism for internal fertilization (Jägersten, 1939b) occurs together with free discharge of the eggs into the water. It is already evident, however, that external fertilization is the original condition for all metazoans. Connected with this mode of fertilization is a special, relatively simple organization of the sperm (the original type of sperm).

3. A small quantity of yolk presupposes free eggs or at least demands that enclosure is not for so long a time that the supply of nutriment for the embryo is endangered. (Here I am disregarding cases of eggs which are secondarily poor in yolk because the mother animal supplies the embryo with nutriment in some other way.) In other words, a small quantity of yolk necessitates planktotrophy in the primary larva or, to express this more cautiously, is generally at least combined with planktotrophy. Increase in the quantity of yolk up to a certain level makes planktotrophy unnecessary and is also followed by lecithotrophy in the larva or even by direct development. This latter type of development is usually combined with lengthy enclosure (see p. 227). Increased quantity of yolk is also known to entail alterations both with regard to the cleavage of the egg and to blastula and gastrula. In eggs containing a large amount of yolk gastrulation by typical invagination is often mechanically impossible.

Although eggs rich in yolk occur also in the groups Spongiaria and Cnidaria which are primitive in the majority of their characters, this does not indicate that abundance of yolk is a phenomenon which is original for the metazoans. The original condition must be tiny eggs containing a relatively small quantity of yolk—a condition which has been fairly commonly retained in the majority of phyla. In this connection I should recall the occurrence of isogametes in certain protozoans and algae. In these organisms, too, abundance of yolk in the female gametes must be a secondary phenomenon.

4. Holoblastic and approximately equal cleavage of the egg presupposes a small quantity of yolk. Pronounced unequal and partial cleavage does not arise except in connection with an increase of yolk. For this reason it seems that the quantity of yolk is the direct cause of the variations in the type of cleavage.

A significant alteration in the type of cleavage in turn entails an alteration of the blastula. In extreme cases this alteration may lead to the disappearance of the blastocoel.

There is no doubt that total and equal cleavage is the original condition. In the division of cells the daughter cells are of different sizes only in special (secondary) cases. (Such a special case is represented in the meiotic divisions of the egg.)

5. The occurrence of the coeloblastula is clearly dependent on the type of cleavage in so far that with pronouncedly unequal, and of course also with partial cleavage, the cavity of the coeloblastula is eliminated. And since the type of cleavage depends on the quantity of yolk, the nature of the blastula is thus indirectly dependent on the amount of yolk. Although the blastocoel has undoubtedly been lost in numerous instances, the coeloblastula is nevertheless still such a common phenomenon, occurring all over the phylogenetic tree, that there is good reason to suppose that an embryonic stage of this kind is original for the metazoans.

It is of great interest to find that certain blastulae (particularly those derived from eggs freely discharged into the water, but also from eggs which are enclosed for a short time) are free-swimming, moving by means of a ciliation which usually covers the whole body. Many non-swimming blastulae are ciliated inside the egg membrane. (In these cases hatching only takes place after gastrulation.)

In my opinion these features indicate clearly that the primeval metazoans possessed a ciliated, free-swimming blastula. This is in perfect agreement with the *Gastraea* theory.

6. Gastrulation by invagination presupposes a blastula with a reasonably spacious blastocoel. Thus alteration of this original blastula-type always involves altered gastrulation. Altered gastrulation (immigration of isolated cells, delamination) occurs occasionally, for example, in Hydrozoa, also in blastulae with a relatively spacious cavity. Irrespective of the explanation of this (the spacious blastocoel might in these cases be a secondary phenomenon) one must not be led by the aberrant conditions in Hydrozoa to the assumption that invagination was not the original mode

of gastrulation. I have referred to this matter already on previous occasions (Jägersten, 1955, 1959).

A rough idea of the wide distribution of the invagination gastrula among the metazoans can be obtained from Fig. 57. This wide distribution, however, is not the only evidence for the primitive nature of invagination. My opinion in this matter is further strengthened by the fact that in several cases where the invagination gastrula is missing, its absence can be explained quite easily by the large quantity of yolk in the egg (cf. above). More or less distinct traces of invagination, in spite of a great mass of yolk occur, moreover, in a number of cases (see e.g. p. 20).

7. Planktotrophy in the pelagic larva is always correlated with small eggs, relatively poor in yolk. Lecithotrophy appears as soon as the quantity of yolk has risen to a certain level which is not necessarily the same in all forms. With lecithotrophy total enclosure and direct development are made possible (for more details, see p. 227). It would be rash to say that an ample supply of yolk *causes* lecithotrophy. It is, however, evident that a certain quantity is a *necessary condition*.

In nearly all phyla with a pelago-benthic life cycle planktotrophic primary larvae are found. As can be seen from Fig. 57, there is also practically full congruence between the distribution of planktotrophy and of invagination gastrula. This is without doubt a result of the relationship of these two phenomena with the quantity of yolk in the egg.

Perhaps the most important general reason for my idea that planktotrophy in the primary larva is an original phenomenon in the metazoans is this relationship to the primitive feature of small eggs, poor in yolk. It is further supported by the observation that as a rule a lecithotrophic larva has been a transitional stage in the various lines of evolution from planktotrophic larva to entirely enclosed direct development. It is not possible to advocate the opposite course (cf. p. 216).

Finally it ought to be pointed out that in the discussion the Spongiaria have been completely omitted. In this group only lecithotrophic larvae are found, and there are particular reasons which make it probable that in the evolutionary line leading to the sponges planktotrophy has never existed (see pp. 12 and 224).

From the above it can be seen that the different phenomena in each of the two ontogenetic systems are more or less intimately interconnected. The central rôle played by the quantity of yolk is striking. An apparent increase in its amount is followed by more or less radical alterations in the majority of the other phenomena, and the primary system is then changed into the secondary system.

It is of interest to note that just the interrelation by itself provides further support for the idea that all the listed phenomena in the primary system are really original.

The ventral band of cilia and its relation to the original blastopore

It would be outside the scope of this book to take up different ancient adult characters for special examination and to attempt comparisons between the phyla on this basis. This must be done in connection with investigations into the evolution of the phyla, and I hope to return to the problem on a later occasion. However, I do want to take the opportunity here to comment on an ancient adult character of more general interest: the longitudinal ventral band of cilia. This occurs in primary larvae of several widely-separated phyla.

The ventral band of cilia is very distinctly developed in forms within the Annelida, where it is inappropriately called "neurotroch" (Fig. 44), Mollusca (Fig. 30), Brachiopoda (Fig. 9), and Enteropneusta (Fig. 53B). In Entoprocta, Sipunculida, and Pogonophora we find a somewhat broader but nevertheless elongated ciliated area which at least in part is a direct counterpart of the band (see below).

In all the groups mentioned the cilia of the band are rather short, usually (perhaps with the exception of the Brachiopoda) considerably shorter than those which serve for swimming in the free water and which as a rule belong to rings of different kinds also occurring in these larvae (pelagic adaptations). The longitudinal band, on the other hand, plays no part, or at most a very subordinate rôle in swimming. It functions instead when the larva creeps over a solid substratum, as can be clearly seen in larvae kept in dishes. This applies to all groups with the possible exception of the Enteropneusta. There, in the *tornaria*-larva, I have not been able to observe locomotion of this kind, but as in the other groups I have seen that particles which come into contact with the band are transported backwards. (These may be useless particles which are rejected from the mouth during feeding. Such a rejection appears not only in the *tornaria*, but also in the larvae of certain entoprocts; see Jägersten, 1964.)

While noting that the band of cilia still has a function at least in the majority of these groups, I want to point out at the same time that this may not apply to all larvae of polychaets in which the band is found. This relinquishment of the band as a creeping organ is undoubtedly the first step in the evolution which leads to its disappearance. In many forms it is already eliminated.

In the cases where mouth and anus are developed, the band of cilia as a rule occupies the greater part of the distance between these openings. It

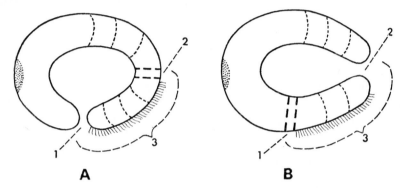

FIG. 59. Diagrams showing the basic organization of the larval body in (A) Annelida and (B) Enteropneusta. 1, mouth; 2, anus; 3, the original blastopore the extension of which in the median plan coincides on the whole with the longitudinal ventral band of cilia occurring in different phyla.

is situated exactly in the suture (at least in certain annelids this is marked by an incision) which is formed in the protostomians, when the blastopore becomes fused from behind during embryogenesis. In the deuterostomian group Enteropneusta, on the other hand, the band is situated in front of the opening which is generally called "blastopore" and which in the course of the development becomes the anus. Yet this opening must only be a remnant of the original blastopore, which in phylogeny has been constricted from in front, and thus only corresponds with the posterior part of the blastopore of the protostomians (prior to its fusion). There is therefore no essential difference between protostomians and deuterostomians with regard to the relation of this band to the *original* blastopore of the coelomates (Fig. 59).

In this connection it might be of interest to note that in more primitive chordates the blastopore is still constricted from in front backwards.

Mouth and anus, respectively, must be considered homologous in proto- and deuterostomians and to have arisen phylogenetically through coalescence of the middle portion of the primeval mouth (mouth-anus) (see Jägersten, 1955). This conception is in fact supported by the existence and extension of the ventral band of cilia in the two main branches of the coelomates. There is reason to suppose that there existed in the adult of the common ancestral form of proto- and deuterostomians (thus before coalescence) a line of cilia with locomotory function ("creeping cilia") on either side of the elongated primeval mouth. After the coalescence of the middle portion of this opening these cilia became situated in median line and in this location formed the unpaired band. This became accelerated to the larva and is still found there (but as a rule no longer in the adult; cf. the shell of the nudibranchiates) in the groups of both protostomians and deuterostomians.

A similar band is also found in the remarkable *zoanthella*-larva belonging to the Anthozoa (see p. 17). I do not consider it impossible that this might be a homologous structure.

Let us examine the conditions mentioned, by reference to Fig. 59. There the changes during the embryogenesis almost up to the fully-developed primary larva in Annelida and Enteropneusta are represented by greatly simplified diagrams. It can be seen that the same basic organization is reached in different ways in a *trochophora* (A) and a *tornaria* (B).

In A the organization of the larva is reached by uneven growth mainly of the blastopore margin, the stronger growth taking place dorsally, and by coalescence of the blastopore from behind forwards, so that finally only a small remnant is left, becoming the definite mouth. (Often this remnant is also closed, and then the mouth is formed later by a perforation in the same place.) The anus is opened at the posterior end.

In B the final organization is reached in another way. The change of shape is the same both dorsally and ventrally, and at first results only in the development of the anus as a remnant of the blastopore at the posterior end. (Sometimes even this opening is temporarily closed.) Later the mouth is formed by a perforation in the anterior part of the ventral side.

Neither of these embryogeneses completely recapitulates phylogeny. Both mouth and anus arose phylogenetically as persistent relics of the anterior and posterior parts of the primeval mouth after the coalescence of its middle portion.

We have to assume that originally the entire digestive sac was a purely adult character (see p. 224), i.e. its development took place only after the transition of the individual to benthic life. Later on, under the influence of adult pressure, the ontogenetic emergence was shifted to the pelagic phase. The same applied to the ventral band of cilia, but not necessarily at the same time.

In the evolutionary period immediately after the appearance of the digestive sac and its openings the ontogeny of these structures without doubt recapitulated their phylogeny. It is difficult to tell why the ontogeny was altered later on to the present course, different in proto- and deuterostomians. However, it is of great importance that the ventral band of cilia makes it possible for us to understand the original extension of the blastopore.

The ventral band of cilia in the larvae is a relic of an extremely ancient organ of locomotion, in all probability the first that was formed in the adult of the metazoans for locomotion on the substratum. In this connection I want to recall the fact that three groups among the protostomians—Entoprocta, Sipunculida, and Mollusca—have a distinctly delimited ventral creeping lobe ("foot") between mouth and anus. The ciliation on this lobe can be considered at least partially homologous with the ventral band. In the larvae of certain gastropods the band has been found to develop before the appearance of the foot, and in later stages the band can occasionally be distinguished in the general ciliation of the lower surface of the foot (Fig. 30B).

It is still premature to form a definite opinion on whether the creeping lobe is one and the same (homologous) structure in the three groups discussed, but I do not think it can be excluded that it is an organ which occurred in the ancestral form common to them. It is well known that Entoprocta, Sipunculida, and Mollusca have also other characters in common. I want to draw attention to the fact that these groups are spiralians with all that this implies, including the *trochophora* larva. Subsequently evolved characters—in the first two groups result mainly from the sessile habit—have, however, caused a marked divergence. If the creeping lobe did exist in the common ancestral form, this must find its expression in the phylogenetic tree, presumably as illustrated in Fig. 57. This is a problem which must certainly be kept in mind in future investigations.

It is quite evident that the creeping lobe, wherever it occurs, is considerably younger than the median band of cilia. As can be understood from the above, the median band must have arisen before the separation of the evolutionary lines leading to proto- and deuterostomians.

Phylogenetic significance of larvae and other juvenile forms. The biogenetic "law"

It is evident that the results I have arrived at in the preceding sections of this book affect in a fundamental way the doctrine of recapitulation and on the whole the phylogenetic significance of the juvenile forms. I shall therefore briefly set out here my views upon these fundamental and so frequently debated problems of zoology.

The significance of the pelagic larvae, especially, has provoked very divergent opinions. According to the advocates of an extreme application of the recapitulation theory even the organization of the primary larva should be a reminiscence of the adult morphology of some pelagic ancestral form. In other words, the larva should recapitulate an earlier adult stage (see below, p. 253). This view has met with opposition from many quarters. It is thought that the larvae, being the result of adaptation to a pelagic mode of life—sometimes it is actually thought that they have arisen only for the purpose of dispersal; see p. 232—cannot recapitulate the *adult* stage of any ancestor. de Beer (1958), especially stands out as an advocate of this critical attitude towards the doctrine of recapitulation. What is recapitulated, or as de Beer says, "repeated" should be only features of the *larvae* of the ancestral forms, not those of their adults. I quite agree with the view that the doctrine of recapitulation must be modified, but this has to be done with careful attention to the fact that the primary larvae can also supply extremely important information about the characteristics of the adult stage of the ancestral forms.

In order not to arrive at erroneous results it must be kept in mind that not all characters in a primary larva have the same status. In the above I have discerned three different kinds:

1. purely larval,
2. recent adult characters,
3. ancient adult characters.

Before making an attempt to interpret the phylogeny of any group a critical examination and evaluation of the different characters in its primary larva is of great importance. In certain cases the evaluation presents no difficulties, in others it involves a varying degree of uncertainty. But this is the only possible way. In the treatment of the different groups I have for this reason attempted to classify the more prominent of the characters of the primary larvae, and have then examined to what extent they can be considered capable of providing information about phylogeny.

Since I have arrived at the opinion that the pelago-benthic life cycle is the original one for all metazoan phyla, and that larva and adult have consequently been adapted to their respective biotopes right from the earliest stage in the metazoan evolution, I must reject the doctrine of recapitulation in its extreme form. I am thus of the opinion that the *purely larval* (*pelagic*) *characters* of the primary larvae tell us nothing about ancestral forms in the *adult* stage. For example, the velum of a molluscan larva or the apparatus of ciliated bands in the *tornaria* of the enteropneusts has never existed in the adult of any ancestor. Neither can the pelagic occurrence of the primary larvae be taken to prove that the evolutionary lines of the various phyla have ever contained pelagic adults. We are thus unable to follow MacBride in his textbook of embryology (1914), in considering that the *pilidium* larva represents the adult stage of a pelagic ancestral form of the Nemertini. It is time that the phylogeneticists relinquished this method of reasoning.

According to a widely held opinion, to which I subscribe without reservation, the recent primary larvae recapitulate the larvae of the ancestral forms farther back in time, often as far back as the common ancestor of two or more phyla (e.g. the *trochophora* groups).

I must therefore stand in opposition to some authors, most of them not phylogeneticists, who maintain that larvae as pelagically adapted forms are unimportant in considerations of phylogeny. This attitude is not correct even in such cases where the larvae lack the ancient adult characters which are so important in this connection. It is perfectly correct to say that the larvae are adapted to their biotopes, but this applies also to the adults, and nobody would suggest that the adults are of doubtful value in phylogeny. I consider it superfluous to enter more closely into the preposterous consequences of the objection outlined. I only want to stress that a fully developed primary larva represents the termination of the first of the two main phases still making up the life cycle in most groups of animals. Often a larva is as valuable for phylogenetic and taxonomic investigations as the adult and much more valuable in the cases where the adult is highly specialized and has recently been altered. In phylogenetic investigations an uncritical use of the larval characters is just as inadmissible as a similar use of the characters of the adults. The features have first to be carefully analysed and evaluated.

It is highly significant that one and the same type of primary larva, such as the *trochophora*, can be found in several otherwise very dissimilar phyla. (This fact has in general received due attention, when phylogenetic and taxonomic attitudes were established.) The fact is also interesting in so far as it fully agrees with, and indeed is an expression of the by no means rare phenomenon that larvae are more resistant to progressive alterations than adults.* The view occasionally expressed, that one and the same larval type, e.g. the *trochophora*, has arisen independently in several different places, has never been very convincing (see e.g. pp. 140 and 143). In the case of pelagic larvae there are admittedly cases of convergence and parallelism in the evolution (see e.g. pp. 198 and 141), but the same conditions are known to exist also in the adults. We can for instance think of all the crowns of tentacles which have arisen independently. But nobody will suggest that for this reason we disregard the adults in phylogenetic investigations. *The only correct procedure is of course the critical and methodical examination of all stages in the life cycle.*

It is important to point out that the morphology of the larvae is by no means as irregular or haphazard as it may appear at first glance. Many groups have their special larval form, and in cases where two or more have one and the same type this sometimes provides the most important proof of a relatively late common ancestor. I might mention as examples Annelida–Mollusca and Enteropneusta–Echinodermata.

It is admittedly a fact that in certain cases quite radical differences exist between the larvae *within* one and the same group, but this previously bewildering situation has in several cases now been fully explained by the principle of adultation. This applies for instance to the in part very dissimilar larvae of the tentaculates (*cyphonautes*, *actinotrocha*, the different larvae of brachiopods) of which some possess certain adult characters that are missing in others, with great differences in the appearance as the result.

Even if, as now established, purely larval (pelagic) features can be of considerable taxonomic-phylogenetic importance, the ancient adult characters which occur in the larvae are still more important. The reason is that they, in contrast to the former type of features, tell us that there

---

* By "progressive" alterations I imply changes which are *new* in the life cycle, and not alterations produced in the larva by adult pressure. In the latter case the larva may admittedly undergo a rapid change, but it is then only a case of displacement within the life cycle of characters which already existed, or possibly of the reduction of larval characters.

have existed ancestral forms somewhere in the line of evolution which have had the same characters in the adult phase of the life cycle. Thus we have a real, although usually incomplete (see below) recapitulation.

The reliability of this method of investigating phylogeny is proved in those cases where the form or group with ancient adult characters has living relatives in which the adult shows the same features. An example may help to illustrate this.

The shell of the *veliger*-larva of the nudibranchiates cannot be a genuine larval (pelagic) character (cf. p. 119). Since, furthermore, the shell is missing in the recent adult, it must in this group be considered as an ancient adult feature. This is shown to be correct in this particular case by the fact that the nudibranchiates have relatives in which the adult still carries a shell. However, according to the principles expressed in this book, the same conclusion should have been reached if all gastropods with shells had become extinct and not left any fossil remains.

Let us now apply the method to the bryozoans. In the *cyphonautes* larva, without doubt the most primitive within this group, well-developed shell structures are likewise found, and their genuine larval nature is no more probable than that of the shell in the nudibranchiate larvae. Neither are these shells found in the recent adult. We are led to the conclusion that here also we have the recapitulation of a character of the adult stage of an ancestral form. The only difference compared with the nudibranchiates is that here there is no other possibility of verifying our conclusion. Because of the great age of the bryozoans and other tentaculates it is unfortunately unlikely that palaeontology will ever help in this matter.

The phylogenetic method outlined is based on two phenomena, firstly adultation and secondly the fact that earlier ontogenetic stages—in these cases the primary larvae—are often less prone to alterations than the adult stages. They are as a rule more conservative than the adults even when it comes to the retention of adult characters. This is the reason for the presence of ancient adult characters in recent larvae.

As has been stated above larvae with ancient adult characters recapitulate only incompletely the features of the adult in some ancestral form, for several reasons. First of all, the adult features are always poorly developed in a larva; it is of course only in the adult that they reach full development. Secondly, they may be modified so as to fit, or at least not to hinder the pelagic mode of life (see p. 223). The ancient adult characters, furthermore, are often reduced as the result of adult pressure (see p. 222).

Taking into account these reservations we can formulate the general statement that *the biogenetic "law" is applicable only in those cases where ancient adult characters are found in a juvenile form*. The type of juvenile form is immaterial. It may be a primary larva in the original pelago-benthic life cycle, a juvenile in a cycle with direct development, or a secondary larva.

It might also be stressed here that in the above discussion I have never maintained that the larvae recapitulate any adult *stages* in phylogeny but only *features* of ancient adults. It seems to me self-evident that a distinction must be made. I have not, however, formed a definite opinion whether or not under exceptional circumstances recapitulation of a whole stage might not occur. I am thinking here of possible cases of evolution via hypermorphosis (prolongation, anaboly), cases where at the end of a life cycle with direct development an addition has been made with the result that a new adult appeared, while the original adult assumed the nature of a secondary larva. If the sexual organs are disregarded, a larva of this kind might perhaps be considered as recapitulating an ancient adult *stage*. The problem is, however, of only peripheral interest to the principles of this book.

I might point out that in the evaluation of characters I never have made use of the terms "palingenetic" and "caenogenetic" introduced by Haeckel. I have avoided these terms because in the literature there are often divergent opinions on the applicability of the one or the other term. The terms used by me have the advantage that the reader can immediately grasp their import. There is, furthermore, a third type—the recent adult characters of the larvae—which have to be classified by themselves.

These recent adult characters do not require much comment here. Their main importance lies in the fact that they show us the effect of adult pressure. Without taking into consideration recent adult features of the larvae we should not be able to attribute to the ancient adult characters the importance these have been given in this book (cf. my comments above on the shell in the *veliger* of the nudibranchiates and in the *cyphonautes* of the bryozoans).

More about recapitulation. The connection between ontogeny and phylogeny

As has already been mentioned the literature shows great divergence of opinion on the validity of the biogenetic "law" and the relation between ontogeny and phylogeny. Since these are questions which are of absolutely fundamental importance in biology, and because in the treatment of the different phyla I have constantly referred to them, I shall take up certain aspects for further discussion. It is impossible here to deal exhaustively with the literature on the subject. I shall have to restrict myself to a summary of my main conclusions.

Haeckel considered that the hereditary changes have taken place mainly at the end of the ontogeny, i.e. in the adult stage. The result of this would be an addition to the development of the individual, and thus a step in phylogeny. Phylogeny would be the mechanical cause of ontogeny by pressing back one adult stage after another into development. This would supply an explanation of the biogenetic "law". Apart from certain secondary alterations, the different stages of ontogeny might quite simply be said to be remains of ancient adult stages.

This concept has been the object of violent criticism. The opponents maintain quite justifiably that the changes take place not only in the adult stage, but may occur at any stage of the life cycle. However, in their attempts to reject recapitulation they go too far and commit a mistake quite as serious as that of the extreme adherents of recapitulation, who believed they could observe the phenomenon almost everywhere.

Even if it is never possible to say categorically that an ontogenetic stage represents an adult ancestral form, i.e. a complete final stage in an ancient life cycle, ontogeny nevertheless often exhibits more or less distinct characters that once existed in the adult stage of ancestral forms. The persistence of such ancient adult characters in many larvae, other juvenile forms, and even embryos is a fact which I have dealt with above in many instances.

There is thus without doubt such a thing as recapitulation, although Haeckel's explanation of the phenomenon is based upon an essentially erroneous concept. Haeckel is, however, right to a certain extent, when he refers the alterations to the adult. I have given several examples of the first phylogenetic emergence of the adult characters in the benthic adult phase and have maintained that this is in accordance with a general rule (p. 221). I refer here to those adult characters in the pelago-benthic life cycle which are adaptations to life on the bottom (and not characters of a pelagic adult which are adaptations to the pelagic life).

Haeckel is partly right also in saying that in the course of evolution a "depression" has taken place in ontogeny, but the depression generally applies not to any entire adult stages but isolated adult characters which thus appeared earlier and earlier in development. What I call "adultation" is essentially a depression of this kind. In other words, the larvae have become similar to the adults so far as these features are concerned.

Recapitulation (the biogenetic "law") rests on the phenomenon of adultation and the fact that the young retain "depressed" adult characters for a longer or shorter time after these have been lost in the adult.

The principle that the earlier stages of a life cycle are often more conservative than the later ones holds for both indirect and direct development. It may be sufficient to give one example for each case.

In certain parasitic gastropods (*Entoconcha* and others) the adult is altered to such a degree that it could not be placed into the proper group without a knowledge of the larvae, which are on the whole shaped like ordinary gastropod *veliger*-larvae. The alterations have begun, according to the rule, in the post-larval phase. (Here and elsewhere the term "alteration" applies of course to the visible expressions and not to the genome.)

As an example for the direct type of development I chose the well-known occurrence of gill slits in the mammalian embryo. These were adduced by Haeckel as proof of recapitulation, and there is nothing to prevent the use of this term provided it is fully understood. In the life cycle of the fishes the gill slits appear at a relatively early stage. The situation is about the same in the mammals, but here the later part of embryogenesis (and of course the whole subsequent phase of the life cycle) has been profoundly altered. One might say that with regard to the gill slits the embryogenesis of the mammals in its earlier stages has not yet arrived at the same level of alteration as it does later. In other words, the recapitulation is again nothing other than the phenomenon that earlier stages of the life cycle are remarkably disinclined towards alteration. Their conservatism goes so far that they still develop adult characters which have been lost long ago in the adult phase.

Haeckel's biogenetic law is sharply attacked by, among others, Garstang (1922) and de Beer (1958). This criticism is, as already said, justified to the extent that in general ontogeny does not recapitulate ancient adult *stages*. Such a criticism is, however, almost equivalent to breaking down an open door. There is no getting away from the fact that the gill slits—to make use of this example once more—are adult characters, features in adult fishes, both ancient and recent. There is no doubt that in the ontogeny of the mammals the gill slits are recapitulated, but of course only in their essential features. The recapitulation is, however, so distinct that there can remain not the slightest doubt that the early ancestors of the mammals even *as adults* had gill slits, i.e. were some kind of fishes. Then it is of minor importance that the recapitulation is demonstrated not by a fully developed character but by a rudiment (anlage). The fact remains, however, that a character which once existed in the adults of the ancestors but was lost in the adults of the descendants is retained in an easily recognizable shape in the embryogenesis of the latter. This is my interpretation of recapitulation (the biogenetic ''law'').

De Beer considers it important to stress that the features which are recapitulated are embryonic or larval features and not those of the adult. The shell-bearing *veliger* of *Entoconcha*, for example, should thus re-capitulate or rather repeat features such as the occurrence of the shell in the larva of the ancestor and not in its adult. Strictly speaking this is probably the most correct method of expression, but *since the shell of the veliger is by derivation an adult character*, which as a result of ''adult pressure'' has secondarily been transferred to the *veliger*, we can say with certainty that the larva of *Entoconcha* demonstrates that the ancestor also had a shell in the adult stage. It is therefore fully justifiable to say that *this* larva recapitulates an adult character in an ancient life cycle.

From the above it is evident that the biogenetic ''law'' can also be applied to features of primary larvae. This validity is, however, restricted, to stress the point once more, to the ancient adult characters of the larvae, thus to characters which in some ancestor were transferred to the pelagic larval phase under the influence of adult pressure (see p. 217 *et seq.*). Furthermore, the biogenetic ''law'' is valid both for direct development and in life cycles with secondary larva. Here also, however, a restriction is necessary; the ''law'' is of course not applicable in the case of secondary deviations such as embryonic membranes. (Adults with such membranes have never existed.) It is the evaluation of the characters which must deter-mine in which cases recapitulation exists.

According to Haeckel phylogeny was the mechanical cause of ontogeny. His opponents (Garstang, 1922; Nauck, 1931; de Beer, 1958; and others) emphatically reject this opinion. Remane (1960) also agrees with them on this point. At first sight it seems evident that the opponents are right. It appears certain that ''phylogeny is the result of ontogeny instead of being its cause'' (de Beer, 1958, p. 9). Garstang (1922) formulates the matter in the following way: ''Ontogeny does not recapitulate phylogeny; it creates it ''. The same view had already been expressed by Müller (1864). From the first, these statements seem so obviously correct that it is difficult to understand how Haeckel has been able to assert the opposite relationship between the two phenomena.

But what is the true state of affairs? Is the means of expressing the relationship between ontogeny and phylogeny really dependent on Haeckel's idea that adult stages have been depressed and have become ontogentic stages? I do not think so.

In the truly unicellular creatures like a simple flagellate, there is no ontogeny in the strict sense of the term, only a kind of regeneration following cleavage. If such a flagellate gives rise to a colony, then also a simple ontogeny of the kind typical of the metazoans has come into existence.

What, then, has happened? It is a hereditary change, a change which finds expression in the fact that at cleavage the cells no longer separate, but are held together. This change, however, is a small step in phylogeny. If afterwards new steps in phylogeny appear, amounting to further divisions of the cells within the colony and differentiation of various kinds, the ontogeny becomes increasingly complicated. This is what has happened to the metazoans. The ontogeny of the present animals, inclusive of their final (adult) stage, is the result of the sum of all the small steps in phylogeny. (Even the simplifications and other changes which have appeared within the ontogeny later on, are of course of a phylogenetical nature.) According to this reasoning it is thus phylogeny which is the cause of ontogeny and not the opposite as maintained by Garstang and de Beer.

I can express myself in this way without assuming with Haeckel that one adult stage after another turns into a stage of ontogeny. It is, however, best not to maintain a causal nexus between ontogeny and phylogeny in either the one or the other direction, but to say that both are caused by hereditary *alterations*. In other words, mutations manifest themselves in ontogeny as well as in phylogeny.

It is evident that the main problem is a struggle over words. The decisive point lies in the definition of the concepts, and how the terms are differentiated from each other. The terms are undoubtedly of great importance, and the discussion in the literature shows how necessary it is that the expressions should be as correct as possible. One must agree with de Beer, when he says: "The matter is important, for it concerns the kernel of biology—development, evolution, and heredity." Fortunately neither recapitulation nor the other phenomena and problems treated in this book depend upon the way in which the connection between phylogeny and ontogeny is formulated.

Haeckel's views on this connection and his explanation of the phenomenon of recapitulation has unfortunately led certain authors to reject the doctrine of recapitulation as a whole. Furthermore, the *Gastraea* theory as well as the germ-layer theory have been rejected with it. On previous occasions (e.g. Jägersten, 1959, p. 97) I have pointed out that the *Gastraea* theory does not necessarily depend on the biogenetic rule. I am still of this opinion, but find at the same time that the recapitulation theory, in the form presented here, supplies strong support for the assumption that at one time in the evolution of the metazoans an ancestral form existed which even as an adult resembled a gastrula—a *Gastraea*. (Unlike Haeckel's *Gastraea* mine is, however, bilaterally symmetrical; Jägersten, 1955, 1959.) I am convinced that both the recapitulation theory and the *Gastraea* theory will survive in modified form.

References

Abeloos, M. (1956). "Les métamorphoses." Collection Armand Colin, Paris.

Åkesson, B. (1958). A study of the nervous system of the Sipunculideae. *Unders. Öresund.* **38.**

Åkesson, B. (1961a). The development of *Golfingia elongata. Ark. Zool.* **13.**

Åkesson, B. (1961b). Some observations on Pelagosphaera larvae. *Galathea Report,* **5,** Copenhagen.

Åkesson, B. (1962). The embryology of *Tomopteris helgolandica. Acta Zool.* **43.**

Anderson, D. T. (1962). The reproduction and early life histories of the gastropods. *Proc. Linn. Soc. N.S.W.* **87.**

Anderson, D. T. (1965a). The reproduction and early life histories of the gastropods *Notoacmaea petterdi, Chiazacmaea flammea* and *Patelloida alticostata* (fam. Acmaeidae). *Proc. Linn. Soc. N.S.W.* **90.**

Anderson, D. T. (1965b). Further observations on the histories of littoral gastropods. *Proc. Linn. Soc. N.S.W.* **90.**

Anderson, D. T. (1966). The comparative embryology of the Polychaeta. *Acta Zool.* **47.**

Ankel, W. E. (1936). Prosobranchia. *Tierwelt N.- u. Ostsee.* **9.**

Ashworth, J. (1916). Larvae of *Lingula* and *Pelagodiscus. Trans. R. Soc. Edinb.* **51.**

Atkins, D. (1932). The ciliary feeding mechanism of the Entoproct Polyzoa. *Q. Jl microsc. Sci.* **75.**

Atkins, D. (1955). The ciliary feeding mechanism of the Cyphonautes larva. *J. mar. biol. Ass. U.K.* **34.**

Baba, K. (1937). Contribution to the knowledge of a nudibranch, *Okadaia elegans. Jap. J. Zool.* **7.**

Baba, K. (1940). The early development of a solenogastre, *Epimenia verrucosa. Ann. Zool. Japan.* **19.**

Baba, K. (1951). General sketch of the development in a solenogastre, *Epimenia verrucosa. Misc. Rep. Res. Inst. Nat. Resources.* **19–21.**

Balfour, F. M. (1880). Essays on Embryology. *Q. Jl microsc. Sci.*

Barrois, J. (1882). Embryogénie des Bryozoaires. *J. Anat. Physiol.,* Paris. **18.**

Bather, F. A. (1900). The Echinoderma. "A Treatise on Zoology." Vol. 3.

Beauchamp, P. de (1907). Morphologie et variations de l'appareil rotateur. *Archs Zool. exp. gén.* **36.**

Beauchamp, P. de (1909). Recherches sur les rotifères. *Archs Zool. exp. gén.* **40.**

Beauchamp, P. de (1929). Le développement des Gastrotriches. *Bull. Soc. zool. Fr.* **54.**

Beer, G. de (1958). "Embryos and Ancestors." Oxford 1958, 3rd ed.

Beklemischew, W. N. (1958). Grundlagen der vergleichenden Anatomie der Wirbellosen, I. Berlin.

Berg, S. E. (1941). Die Entwicklung und Koloniebildung bei *Funiculina quadrangularis*. *Zool. Bidr. Upps.* **20**.

Berkeley, E. and Berkeley, C. (1963). Neoteny in larvae of two species of Spionidae. *Can. J. Zool.* **41**.

Berrill. N. J. (1955). "The origin of the Vertebrates." Clarendon Press, Oxford.

Bone, Q. (1960). The origin of the Chordates. *J. Linn. Soc. Lond.*, Zoology **44**.

Braem, F. (1890). Untersuchungen über die Bryozoen des süssen Wassers. *Zoologica, Stuttg.* **10**.

Braem, F. (1897). Die geschlechtliche Entwicklung von *Plumatella fungosa*. *Zoologica, Stuttg.* **10**.

Braem, F. (1908). Die geschlechtliche Entwicklung von *Fredericella sultana*. *Zoologica, Stuttg.* **20**.

Braem, F. (1939). *Victorella sibogae. Z. Morph. Ökol. Tiere.* **36**.

Braem, F. (1951). Über *Viktorella* und einige ihrer Verwandten . . . *Zoologica, Stuttg.* **37**.

Brien, P. (1953). Étude sur les Phylactolaemates. *Annls Soc. r. zool. Belg.* **84**.

Brien, P. (1960). Classe des Bryozoaires. "Traité de Zoologie." Vol. 5.

Brien, P. and Papyn, L. (1954). Les entoproctes et la classe des bryozoaires. *Annls Soc. r. zool. Belg.* **85**.

Burdon-Jones, C. (1952). Development and biology of the larva of *Saccoglossus horsti*. *Phil. Trans. R. Soc. Lond.* B **236**.

Cantell, C.-E. (1966a). The devouring of the larval tissues during the metamorphosis of pilidium larvae (Nemertini). *Ark. Zool.* **18**.

Cantell, C.-E. (1966b). Some developmental stages of the peculiar nemertean larva pilidium recurvatum . . . *Ark. Zool.* **19**.

Cantell, C.-E. (1969). Morphology, development and biology of the pilidium larvae . . . *Zool. Bidr. Upps.* **38**.

Carlgren, O. (1924). Die Larven der Ceriantharien, Zoantharien und Actiniarien . . . *Ergebn. deutsch. Tiefsee-Exp.* **19**.

Carter, G. S. (1957). Chordate Phylogeny. *Syst. Zool.* **6**.

Clark, H. L. (1910). The development of an apodous holothurian (*Chiridota rotifera*). *J. exp. zool.* **9**.

Coe, W. R. (1943). Biology of the nemerteans of the Atlantic coast of North America. *Trans. Conn. Acad. Arts Sci.* **35**.

Colvin, A. L. and Colvin, L. H. (1950). The development capacities . . . of *Sacoglossus kowalevskyi*. *J. exp. Zool.* **115**.

Conklin, E. G. (1902). The embryology of a brachiopod, *Terebratulina septentrionalis. Proc. Am. phil. Soc.* V **41**.

Conklin, E. G. (1908). Two peculiar actinian larvae from Tortugas Florida. *Carnegie Inst. Washington*. **103.**

Cori, C. J. (1936). Kamptozoa. "Bronns Klassen und Ordungen des Tierreichs."

Cori, C. J. (1939). Phoronidea. "Bronns Klassen und Ordungen des Tierreichs."

Damas, H. (1962). La collection de pelagosphaera du "Dana". *Dana Rep*. **59.**

Dawydoff, C. (1940a). Les formes larvaires de Polyclades et de Némertes du plancton Indochinois. *Bull. biol. Fr. Belg*. **74.**

Dawydoff, C. (1940b). Quelques véligères géantes de Prosobranches provenant de la Mer de Chine. *Bull. biol. Fr. Belg*. **74.**

Delsman, H. (1915). Eireifung und Gastrulation bei *Emplectonema*. *Tijdschr. ned. dierk. Vereen*. Ser. 2, **14.**

Dieck, G. (1874). Beiträge zur Entwicklungsgeschichte der Nemertinen. *Jena. Z. Naturw*. **8.**

Dodd, J. M. (1957). Artificial fertilization, larval development and metamorphosis in *Patella vulgata* and *Patella coerulea*. *Publ. staz. zool. Napoli*. **29.**

Drew, G. A. (1899). Some observations on the habits, anatomy and embryology of the Protobranchia. *Anat. Anz*. **15.**

Drew, G. A. (1901). Life history of *Nucula delphinodonta*. *Q. Jl microsc. Sci*. **44.**

Edwards, Ch. L. (1908). Variation, development and growth in *Holothuria floridana . . . Biometrika*. **6.**

Fell, H. B. (1941). The direct development of a New Zealand ophiuroid. *Q. Jl microsc. Sci*. **82.**

Fell, H. B. (1948). Echinoderm embryology and the origin of chordates. *Biol. Rev*. **23.**

Fewkes, J. W. (1883). On the development of certain worm larvae. *Bull. Mus. comp. Zool. Harv*. **11.**

Fioroni, P. (1966). Zur Morphologie und Embryogenese des Darmtraktes . . bei Prosobranchiern. *Revue suisse Zool*. **73.**

Fioroni, P. and Sandmeier, E. (1964). Über eine neue Art der Nähreibewältigung bei Prosobranchierveligern. *Vie Milieu (suppl.)* **17.**

Franzén, A. (1955). Comparative morphological investigations into the spermiogenesis among Mollusca. *Zool. Bidr. Upps*. **30.**

Franzén, A. (1956). On spermiogenesis, morphology of the spermatozoon, and biology of fertilization among invertebrates. *Zool. Bidr. Upps*. **31.**

Franzén, A. (1967). A new loxosomatid from the Pacific (Gilbert Islands) with a note on internal budding. *Ark. Zool*. **19.**

Fretter, V. (1943). Studies in the functional morphology and embryology of *Oncidiella*. *J. Mar. biol. Ass. U.K*. **25.**

Friedrich, H. (1949). Lebensformtypen bei pelagischen Polychaeten. *Zool. Anz. (Suppl.)* **13.**

Garstang, W. (1922). The theory of recapitulation. A critical restatement of the biogenetic law. *J. Linn. Soc. Lond.* Zoology **35**.

Garstang, W. (1928a). The origin and evolution of larval forms. *Rep. Br. Ass. Advmt. Sci.* for 1928.

Garstang, W. (1928b). The morphology of the Tunicata, and its bearing on the phylogeny of the Chordata. *Q. Jl microsc. Sci.* **72**.

Garstang, W. (1939). Spolia Bermudiana. *Q. Jl microsc. Sci.* **81**.

Gegenbaur, C. (1855). ''Untersuchungen über Pteropoden und Heteropoden.'' Leipzig.

Geigy, R. and Portmann, A. (1941). Versuch einer morphologischen Ordnung der tierischen Entwicklungsgänge. *Naturwiss.* **29**, 49.

Gerould, J. H. (1904). Studies on the embryology of the Sipunculidae. 1. The embryonal envelope and its homologue. *Mark. Annivers. Vol. New York.* **32**.

Gerould, J. H. (1907). The development of *Phascolosoma*. *Zool. Jb* (Abt. Anat.) **23**.

Grave, C. (1903). On the occurrence among echinoderms of larvae with cilia arranged in transverse rings, with a suggestion as to their significance. *Biol. Bull.* **5**.

Gravier, Ch. (1923). La ponte et l'incubation chez les Annélides Polychètes. *Annls Sci. nat.* **6**.

Hadži, J. (1963). ''The evolution of the Metazoa.'' Pergamon Press, London.

Hadži, J. (1964). Genetic relationship between pelagic and benthal organisms. *Acta adriat.* **11**.

Häcker, V. (1896). Pelagische Polychaetenlarven. *Z. wiss. Zool.* **62**.

Häcker, V. (1898a). Die pelagischen Polychaeten- und Achätenlarven der Plankton-Expedition. *Ergebn. Plankton-Exp.* **2**.

Häcker, V. (1898b). Pelagische Polychaetenlarven. *Biol. Zbl.* **18**.

Hammarsten, O. D. (1915). Zur Entwicklungsgeschichte von *Halicryptus spinulosus*. *Z. wiss. Zool.* **112**.

Hand, C. (1959). On the origin and phylogeny of the coelenterates. *Syst. Zool.* **8**.

Hannerz, L. (1956). Larval development of the polychaete families Spionidae . . . *Zool. Bidrag Uppsala.* **31**.

Harmer, S. F. (1886). On the life-history of *Pedicellina*. *Q. Jl microsc. Sci.* **27**.

Hatschek, B. (1877). Embryonalentwicklung und Knospung der *Pedicellina*. *Z. Wiss. Zool.* **29**.

Hatschek, B. (1881). Entwicklungsgeschichte von *Echiurus*. *Arb. zool. Inst. Univ. Wien.* **3**.

Hatschek, B. (1884). Ueber Entwicklung von *Sipunculus nudus*. *Arb. zool. Inst. Univ. Wien.* **5**.

Hatschek, B. (1885). Entwicklung der Trochophora von *Eupomatus unicinatus*. *Arb. zool. Inst. Univ. Wien.* **6**.

Heath, H. (1899). The development of *Ischnochiton. Zool. Jb.* **12.**

Heath, H. (1918). Solenogastres from the eastern coast of North America. *Mem. Mus. comp. Zool. Harv.* **45.**

Heath, H. (1928). A sexually mature turbellarian resembling Müller's larva. *J. Morph.* **45.**

Heintz, A. (1939). Die Entwicklung des Tierreiches. *Naturwiss.*

Hubendick, B. (1952). *Veloplacenta,* a new genus of prosobranchiate Mollusca. *Ark. Zool.* **3.**

Huus, J. (1931). Über die Begattung bei *Nectonema* und über den Fund der Larve. *Zool. Anz.* **97,** 1932.

Hyman, L. H. (1940). "The Invertebrates: Protozoa through Ctenophora." McGraw-Hill, New York.

Hyman, L. H. (1951a). "The Invertebrates: Plathelminthes and Rhynchocoela." McGraw-Hill, New York.

Hyman, L. H. (1951b). "The Invertebrates: Acanthocephala. Aschelminthes, and Entoprocta." McGraw-Hill, New York.

Hyman, L. H. (1955). "The Invertebrates: Echinodermata." McGraw-Hill, New York.

Hyman, L. H. (1959). "The Invertebrates: Smaller coelomate groups." McGraw-Hill, New York.

Ikeda, I. (1901). Observations on the development, structure, and metamorphosis of *Actinotrocha. J. Coll. Sci. imp. Univ. Tokyo.* **13.**

Ivanov, A. V. (1963). "Pogonophora." Academic Press, London.

Iwanoff, P. P. (1928). Die Entwicklung der larvalsegmente bei den Anneliden. *Z. Morph. Öleol, Tiere.* **10.**

Iwata, F. (1958). On the development of the nemertean *Micrura akkeshiensis. Embryologia.* **4.**

Iwata, F. (1960). Studies on the comparative embryology of nemerteans with special reference to their interrelationships. *Publs Akkeshi mar. biol. Stn.* **10.**

Jurczyk, C. (1927). Beiträge zur Morphologie, Biologie und Regeneration von *Stephanoceros. Z. wiss. Zool.* **129.**

Jägersten, G. (1939a). Zur Kenntnis der Larvenentwicklung bei *Myzostomum. Ark. Zool.* **31A.**

Jägersten, G. (1939b). Über die Morphologie und Physiologie des Geschlechtsapparats und den Kopulationsmechanismus der Myzostomiden. *Zool. Bidr. Upps.* **18.**

Jägersten, G. (1940a). Zur Kenntnis der äusseren Morphologie. Entwicklung und Ökologie von *Protodrilus rubropharyngeus* n. sp. *Ark. Zool.* **32A.**

Jägersten, G. (1940b). Die Abhängigkeit der Metamorphose vom Substrat des Biotops bei *Protodrilus*. *Ark. Zool.* **32A**.

Jägersten, G. (1940c). Zur Kenntnis der Morphologie, Entwicklung und Taxonomie der Myzostomida. *Nova Acta R. Soc. Scient. Upsal.* Ser. 4, **11**.

Jägersten, G. (1944). Zur Kenntnis der Morphologie, Enzystierung und Taxonomie von *Dinophilus*. *K. Sv. Vetensk. Handl.* **21**.

Jägersten, G. (1952). Studies on the morphology, larval development and biology of Protodrilus. *Zool. Bidr. Upps.* **29**.

Jägersten, G. (1955). On the early phylogeny of the Metazoa. The Bilaterogastraea-theory. *Zool. Bidr. Upps.* **30**.

Jägersten, G. (1956). Investigations on *Siboglinum ekmani* n. sp., encountered in Skagerak. *Zool. Bidr. Upps.* **31**.

Jägersten, G. (1957). On the larva of *Siboglinum*. *Zool. Bidr. Upps.* **32**.

Jägersten, G. (1959). Further remarks on the early phylogeny of the Metazoa. *Zool. Bidr. Upps.* **33**.

Jägersten, G. (1963). On the morphology and behaviour of Pelagosphaera larvae. *Zool. Bidr. Upps.* **36**.

Jägersten, G. (1964). On the morphology and reproduction of entoproct larvae. *Zool. Bidr. Upps.* **36**.

Jeschikov, J. (1936). Metamorphose, Cryptometabolie und direkte Entwicklung. *Zool. Anz.* **114**.

Kaester, A. (1963). "Lehrbuch der Speziellen Zoologie." Gustav Fischer, Jena.

Kato, K. (1940). On the development of some Japanese Polyclads. *Jap. J. Zool.* **8**.

Kessel, M. M. (1964). Reproduction and larval development of *Acmaea testudinalis*. *Biol. Bull.* **127**.

Kirk, H. B. (1938). Notes on the breeding habits and early development of *Dolichoglossus otagoensis*. *Trans. Proc. R. Soc. N.Z.* **88**.

Kohn, A. J. (1960). Development in marine gastropod molluscs of the genus *Conus* and its ecological significance. *Anat. Rec.* **138**.

Kohn, A. J. (1961a). Studies on spawning behavior, egg masses, and larval development in the gastropod genus *Conus*. I. Observations on nine species in Hawaii. *Pacif. Sci.* **15**.

Kohn, A. J. (1961b). Spawning behavior, egg masses, and larval development in the gastropod genus *Conus*. II. Observations in the Indian Ocean. *Bull Bingham oceanogr. Coll.* **17**.

Korschelt, E. and Heider, K. (1890). Lehrbuch der vergleichenden Entwicklungsgeschichte, spez. Teil. Jena.

Korschelt, E. and Heider, K. (1936). Vergleichende Entwicklungsgeschichte der Tiere. Jena.

Kupelwieser, H. (1905). Bau und Metamorphose des Cyphonautes. *Zoologica*. **19**.

Lang, A. (1884). Die Polycladen (Seeplanarien) des Golfes von Neapel und der angrenzenden Meeresabschnitte. *Fauna Flora Golf. Neapel* **11**.

Lang, K. (1948). On the morphology of the larva of *Priapulus caudatus*. *Ark. Zool.* **41A**.

Lang, K. (1953). Die Entwicklung des Eies von *Priapulus caudatus*. *Ark. Zool.* **41A**.

Lebour, M. V. (1931). Clione limacina in Plymouth waters. *J. mar. biol. Ass. U.K.* **17**.

Lebour, M. V. (1937). The eggs and larvae of the British prosobranchs. *J. mar. biol. Ass. U.K.* **22**.

Lemche, H. (1963). The origin of the Bryozoa Phylactolaemata. *Vidensk. Medd. Dansk. naturh. Foren.* **125**.

MacBride, E. W. (1914). "Text-book of Embryology. I." London.

Marcus, E. (1938). Bryozoarios marinhos Brasileiros, II. *Bol. Fasc. Fil. Ciên. Letr. Univ. S. Paulo. 4. Zoologica.* **2**.

Marcus, E. (1939). Bryozoarios marinhos Brasileiros, III. *Bol. Fasc. Fil. Ciên. Letr. Univ. S. Paulo. 13, Zoologica.* **3**.

Mariscal, R. N. (1965). The adult and larval morphology and life history of the entoproct *Barentsia gracilis*. *J. Morph.* **116**.

Masterman, A. (1900). On the Diplochorda, 3. The early development and anatomy of *Phoronis buski*. *Q. J. Micr. Sci.* **43**.

Metschnikoff, E. (1884). Vergleichend embryologische Studien. *Z. wiss Zool.* **42**.

Metschnikoff, E. (1886). "Embryologische Studien an Medusen." Vienna.

Mortensen, Th. (1921). Studies of the development and larval forms of echinoderms. Copenhagen.

Mortensen, Th. (1923). Echinoderm larvae and their bearing on classification. *Nature*, **111**.

Mortensen, Th. (1931). Contributions to the study of the development and larval forms of echinoderms, I, II. *K. Dansk Vidensk. Selsk. Skr. Nat. Math. Avd.* **4**.

Mortensen, Th. (1937). Contributions to the study of the development and larval forms of echinoderms, III. *K. Dansk. Vidensk. Selsk. Skr. Nat. Math. Avd.* **7**.

Mühldorf, A. (1914). Beiträge zur Entwicklungsgeschichte der *Gordius*-Larve. *Z. wiss. Zool.* **111**.

Müller, F. (1864). "Für Darwin." Leipzig.

Nauck, E. T. (1931). Über umwegige Entwicklung. *Morph. Jahrbuch.* **66**.

Nielsen, C. (1966). On the life-cycle of some Loxosomatidae. *Ophelia.* **3**.

Nielsen, C. (1967). Metamorphosis of the larva of *Loxosoma murmanica*. *Ophelia.* **4**.

Nørrevang, A. (1964). Choanocytes in the skin of *Harrimannia* (Enteropneusta). *Nature.* **204.**

Nyholm, K.-G. (1943). Zur Entwicklungsbiologie der Ceriantharien und Actinien. *Zool. Bidrag Uppsala.* **22.**

Nyholm, K.-G. (1947). Studies in the Echinoderida. *Ark. Zool.* **39A.**

Nyholm, K.-G. (1949). On the development and dispersal of Athenaria Actinia with special reference to *Halcampa duodecimcirrata. Zool. Bidr. Upps.* **27.**

Nyholm, K.-G. (1951). The development and larval form of *Labidoplax buskii, Zool. Bidr. Upps.* **29.**

Ohshima, H. (1921). On the development of *Cucumaria echinata. Q. J. Microsc. Sci.* **65.**

Ohshima, H. (1925). Notes on the development of the sea-cucumber *Thyone briareus. Science.* **61.**

Orrhage, L. (1966). Über die Anatomie des zentralen Nervensystemes der sedentären Polychaeten. *Ark. Zool.* **19.**

Percival, E. (1944). A contribution to the life-history of the brachiopod *Terebratella inconspicua. Trans. Roy. Soc. N.Z.* **74.**

Percival, E. (1960). A contribution to the life-history of the brachiopod *Tegulorhynchia nigricans. Q. J. microsc. Sci.* **101.**

Pierantoni, U. (1908). Protodrilus. *Fauna u. Flora des Golfes von Neapel.* **31.**

Plenk, H. (1915). Die Entwicklung von *Cistella. Arb. Zool. Inst. Wien.* **20.**

Portmann, A. (1955). La métamorphose "abritée" de *Fusus. Rev. Suisse Zool.* **62,** *fasc. suppl.*

Pruvot, G. (1890). Sur le developpement d'un solénogastre. *C. R. Acad. Sci. Paris.* **111.**

Pruvot, G. (1892). Sur l'embryogénie d'une *Proneomenia. C. R. Acad. Sci. Paris.* **114.**

Purasjoki, K. J. (1944). Beiträge zur Kenntnis der Entwicklung und Ökologie der *Halicryptus spinulosus*-Larve. *Ann. Zool. Soc. Zool. Bot. Fennicae Vanamo.* **9.**

Rasmussen, E. (1956). Faunistic and biological notes on marine invertebrates. *Biol. Medd. Dansk. Vid. Selsk.* **23.**

Reinhardt. H. (1941). Entwicklungsgeschichte der *Prostoma. Vierteljahrschr. Nat. Gesellsch. Zürich.* **86.**

Remane, A. (1956). Zur Biologie des Jugendstadiums der Ctenophore *Pleurobrachia pileus. Kieler Meeresforsch.* **12.**

Remane, A. (1960). Die Beziehungen zwischen Phylogenie und Ontogenie. *Zool. Anz.* **164.**

Remane, A. (1963). Zur Metamerie, Metamerismen und Metamerisation bei Wirbeltieren. *Zool. Anz.* **170.**

Riedl, R. (1959). Beiträge zur Kenntnis der *Rhodope veranii*, I, Geschichte und Biologie. *Zool. Anz.* **163**.

Riedl, R. (1960). Beiträge zur Kenntnis der *Rhodope veranii*. II, Entwicklung. *Zeitschr. wiss. Zool.* **163**.

Rowell, A. J. (1960). Some early stages in the development of the Brachiopod *Crania anomala* (Müller). *Ann. Mag. Nat. Hist.* **3**, ser. **13**.

Rullier, F. (1955). Développement du Serpulien *Mercierella enigmatica* Fauvel. *Vie. et Milieu.* **4**, 1955.

Runnström, J. and S. (1921). Entwicklung von *Cucumaria frondosa* und *Psolus phantapus*. *Bergens. Mus. Aarbok, 1918–19. Naturvid. Rekke.* **5**.

Sachwatkin, A. A. (1956). "Vergleichende Embryologie der niederen Wirbellosen." Berlin.

Scheltema, R. S. (1962). Pelagic larvae of New England intertidal gastropods. *Trans. Amer. micr. soc.* **81**.

Schepotieff, A. (1909). Die Pterobranchier des Indischen Ozeans. *Zool. Jahrb. Abt. Syst.* **28**.

Schmidt, G. A. (1934). Ein zweiter Entwicklungstypus von *Lineus gesserensis-ruber. Zool. Jahrb. Abt. Anat.* **58**.

Schmidt, G. A. (1966). "Evolutionäre Ontogenie der Tiere." Berlin.

Schultz, E. (1897). Über Mesodermbildung bei *Phoronis. Trav. Soc. Nat. St. Pétersbourg.* **28**.

Schultz, E. (1910). Prinzipen der rationalen vergleichenden Embryologie. Leipzig.

Segrove, F. (1938). Surface ciliation in some polychaete worms. *Proc. Zool. Soc. Lond. B.* **108**.

Segrove, F. (1941). The development of the serpulid *Pomatoceros triqueter* Q. *J. Microsc. Sci.* **82**.

Selys-Longchamps, M. de. (1907). *Phoronis. Fauna u. Flora des Golfes von Neapel.* **30**.

Semon, R. (1888). Die Entwicklung der *Synapta digitata* und die Stammesge-schichte der Echinodermen. *Jenaische Zeitschr.* **22**.

Shearer, A. (1911). Development and structure of the trochophore of *Hydroides*. *Q. J. Microsc. Sci.* **56**.

Silén, L. (1944a). The anatomy of Labiostomella gisleni. With special regard to the embryo chambers . . . *K. Sv. Vetenskapsakad. Handl.* **21**.

Silén, L. (1944b). The main features of the development of . . . Bryozoa Gymnolaemata. *Ark. Zool.* **35A**.

Silén, L. (1954). Developmental biology of the Phoronidea of the Gullmar fiord area. *Acta Zoologica.* **35**.

Smith, J. E. (1935). Early development of *Cephalothrix. Q. J. Microsc. Sci.* **77**.

Spengel, J. W. (1893). Die Enteropneusten des Golfes von Neapel. *Fauna u. Flora des Golfes von Neapel*. **18**.

Sterrer, W. (1966). New polylithophorous marine Turbellaria. *Nature*. **210**.

Surface, F. M. (1907). The early development of a polyclad, *Planocera inquilina*. *Proc. Acad. Nat. Sci. Philadelphia*. **59**.

Swedmark, B. (1954). Etude du développement larvaire et remarques sur la morphologie de *Protodrilus symbioticus*. *Ark. Zool*. **6**.

Sveshnikov, V. A. (1960). Pelagic larvae of some Polychaeta in the White Sea. Зоологический Журнал. **39**.

Söderström, A. (1924a). Katastrophale Metamorphose der *Polygordius*-Endolarve und Spiralfurchung. *Uppsala Univ. Årsskr. (Mat. Nat.)*. 1924.

Söderström, A. (1924b). "Das Problem der *Polygordius*-Endolarve." Uppsala.

Söderström, A. (1925a). Das Problem der *Polygordius*-Endolarve. *Zool. Anz*. **64**.

Söderström, A. (1925b). "Die Verwandtschaftsbeziehungen der Mollusken." Uppsala.

Thompson, T. E. (1959a). Feeding in nudibranch larvae. *J. mar. biol. Ass. U.K*. **38**.

Thompson, T. E. (1959b). Development of the aplacophorous mollusc *Neomenia carinata* Tullberg. *Nature*. **184**.

Thompson, T. E. (1960). The Development of *Neomenia carinata* Tullberg. *Proc. Roy. Soc. B*. **153**.

Thompson, T. E. (1967). Direct development in a nudibranch, *Cadlina laevis*, with a discussion of developmental processes in Opisthobranchia. *J. mar. biol. Ass. U.K*. **47**.

Thorson, G. (1940). Studies on the egg masses and larval development of Gastropoda from the Iranian Gulf. *Danish Sc. Invest. in Iran*, II. Copenhagen.

Thorson, G. (1946). Reproduction and larval development of Danish marine bottom invertebrates. *Meddel. Komm. Danm. Fisk.- og Havunders*. **4**, Ser. Plankton.

Thorson, G. (1951). Zur jetzigen Lage der marinen Bodentier-Ökologie. *Verhandl. Deutsch. Zool. Gesellsch*.

Webb, M. (1964). The larvae of *Siboglinum fiordicum* . . . *Sarsia*. **15**.

Webb, M. (1965). Additional notes on the adult and larva of *Siboglinum* etc. *Sarsia*. **20**.

Werner, B. (1955). Über die Anatomie, die Entwicklung und Biologie des Veligers und der Veliconcha von *Crepidula fornicata*. *Helgoland. wiss. Meeresunters*. **5**.

Westblad, F. (1937). Die Turbellarien-Gattung *Nemertoderma*. *Acta Soc. Fauna Flora Fennica*. **60**.

Westblad, F. (1949). On *Meara stichopi* (Bock) Westblad, a new representative of Turbellaria archoophora. *Ark. Zool.* **1**.

Whitear, M. (1957). Some remarks on the Ascidian affinities of vertebrates. *Ann. Mag. nat. Hist.* **10**.

Widersten, B. (1968). On the morphology and development in some Cnidarian larvae. *Zool. Bidrag Uppsala.* **37**.

Wietrzykowski, W. (1914). Recherches sur de développement de *l'Edwardsia beautempsii*. *Bull. Int. Acad. Sc. Cracovie.* 1914 B.

Wilson, D. P. (1932). On the Mitraria-larva of *Owenia fusiformis*. *Philos. Trans. R. Soc. Lond.* B. **221**.

Wilson, D. P. (1936). The development of the sabellid *Branchiomma vesiculosum*. *Q. J. microsc. Sci.* **78**.

Woltereck, R. (1902). Ueber zwei Entwickelungstypen der *Polygordius*-Larve. *Verhandl. des I. Internationalen Zoologen-Congresses zu Berlin.*

Woltereck, R. (1905). Wurmkopf. Wurmrumpf und Trochophora. *Zool. Anz.* **28**.

Woltereck, R. (1925). Die *Polygordius*-Endolarve. *Zool. Anz.* **63**.

Yatsu, N. (1902). Development of *Lingula*. *J. Coll. Sci. imp. Univ. Tokyo.* **17**.

Zimmer, R. L. (1964). Reproductive biology and development of Phoronida. *Dissertation Abstracts.* **25**. 1965.

Author Index

Subject Index